Silver Link Silk Editions

The Steam Engines of World War II in Europe

Silver Link Books

WEST GERMANY: Trains using the line which ran down the Mosel Valley from Koblenz to Trier remained steam-hauled into the 1970s. Freight traffic was headed by ex-DRG 2-10-0s while Class 23 2-6-2s, first introduced in 1950, were used on passenger trains. Here DB Class 44 2-10-0 No 044.331-7 approaches Cochem with a southbound freight in early September 1970. No 044.331-7 (ex-DRG 44.1331 (Krupp, No 2753)) was built as an 'Übergangskriegslokomotive' or Transitional War Locomotive (ÜK) in 1942 and was withdrawn in June 1974 *John Price*

The Steam Engines of World War II in Europe

Philip Horton

Silver Link Books

© Philip Horton 2021

All rights reserved. No part of this publication may be reproduced, stored in a retrieval system or transmitted, in any form or by any means, electronic, mechanical, photocopying, recording or otherwise, without prior permission in writing from Silver Link Books, Mortons Media Group Ltd.

First published in 2021

British Library Cataloguing in Publication Data

A catalogue record for this book is available from the British Library.

ISBN 978 1 85794 569 0

Silver Link Books
Mortons Media Group Limited
Media Centre
Morton Way
Horncastle
LN9 6JR
Tel/Fax: 01507 529535

email: sohara@mortons.co.uk
Website: www.nostalgiacollection.com

Printed and bound in the Czech Republic

Acknowledgements

As I was born in 1947 my own encounters with the steam engines of the Second World War have involved only their post-war work and preservation. In writing this book I have therefore had to draw on a great many sources of information, and these are listed in Appendix 3. I would particularly like to thank Mortons Media Group, who have granted me permission to reproduce material from wartime editions of *The Railway Magazine*, which they now publish. Further information about the Group's publications can be obtained from its website: www.classicmagazines.co.uk.

Much information and many photographs have been provided by friends and associates, many of whom, like me, pursued steam into Western, then Eastern Europe after 1968. These include Brian May, John Price, David Alison, Adrian Cornish, David Bate, Steve Frost, Barry Payne, Steve Roberts and Frank Hornby. Slip Coach Publishing Services have also kindly allowed me to use photos from the collection of the late Ray Ruffell and Terry Gough. Other photos have been provided courtesy of Gregor Atzbach via his website, www.bundesbahn.net, the South Limburg SteamTrain Company, Reinier Zondervan, SteamTrain Goes-Borsele and the Dutch National Railway Museum in Utrecht. The photo of the 'Kriegslok Sculpture' is courtesy of Masur via Wikipedia Commons. I must also especially thank Stephen Edge, who has contributed the maps, as he has for all my previous publications. All uncredited photos are my own.

Finally I must again thank my wife Sue for accompanying me around Europe over the last five decades, and also for acting as my translator. She has also helped me proofread and edit the text. The encouragement received from Peter Townsend of Slip Coach Publishing and Will Adams of Keyword is also acknowledged.

Philip Horton
Lincolnshire
October 2020

Contents

Map 1 European country boundaries post 1919 — 6
Map 2 Changes in European country boundaries after 1945 and 1994 — 7
General introduction — 8

Part 1 Kriegsdampflokomotiven – German War Engines

Introduction — 9

1. The German Übergangskriegslokomotiven – Transitional War Locomotives (ÜK)
1.1 The DRG Class 86 2-8-2 tank engines — 10
1.2 The DRG Class 44 2-10-0s — 12
1.3 The DRG Class 50 2-10-0s — 15

2. Kriegsdampflokomotiven – War Steam Locomotives (KDL)
2.1 The DRG Class 52 'Austerity' 2-10-0s (KDL 1) — 23
2.2 The DRG Class 42 'Austerity' 2-10-0s (KDL 3) — 48
2.3 Other KDL and Heeresfeldbahnlokomotiven (HF) – German Army Locomotives — 50

Part 2 British and American War Engines

Introduction to Chapters 3 and 4 — 60

3. British War Department Steam Engines
3.1 The Stanier 8F 2-8-0s — 62
3.2 The Hunslet 'Austerity' 0-6-0 saddle tank engines — 74
3.3 The WD 'Austerity' 2-8-0s — 88
3.4 The WD 'Austerity' 2-10-0s — 95
3.5 Bulleid's Southern Railway 'Austerity' 0-6-0s — 103

4. American War Engines
4.1 The USATC Class S100 0-6-0 tank engines — 107
4.2 The USATC Class S160 2-8-0s — 113

Appendix 1 Engine numbering on Europe's railways, 1939-90 — 127
Appendix 2 Abbreviations used in the text — 128
Appendix 3 Bibliography and other references — 131

Table 1 Selected ex-DRG Class 86, 44, 50, 52 and 42 engines preserved in Germany — 132
Table 2 Selected ex-DRG Class 86, 44, 50, 52 and 42 engines preserved outside Germany — 137
Table 3 Other selected KDL and HF engines preserved in Great Britain and Europe — 149
Table 4 Selected WD engines preserved in Great Britain and Europe — 151
Table 4a Other Stanier 8Fs and 'Austerity' 0-6-0 saddle tanks preserved in Great Britain — 155
Table 5 Selected USATC engines preserved in Great Britain and Europe — 157
Table 5a USATC designed engines, now preserved in Great Britain — 159

The Steam Engines of World War II in Europe

Map 1: European country boundaries post 1919

During the period covered by this book numerous changes occurred to the boundaries of European countries, particularly in the east. The situation immediately before the German invasion of Czechoslovakia in 1939 reflected the boundaries imposed on Germany, Austria and Hungary by the various treaties following the end of World War I. Germany and Austria had lost extensive territory through the formation of the newly independent countries of Poland, Czechoslovakia and Jugoslavia, while Hungary had lost a swathe of its land in the east to Romania. Belarus and Ukraine were briefly independent states before their absorption into the USSR.

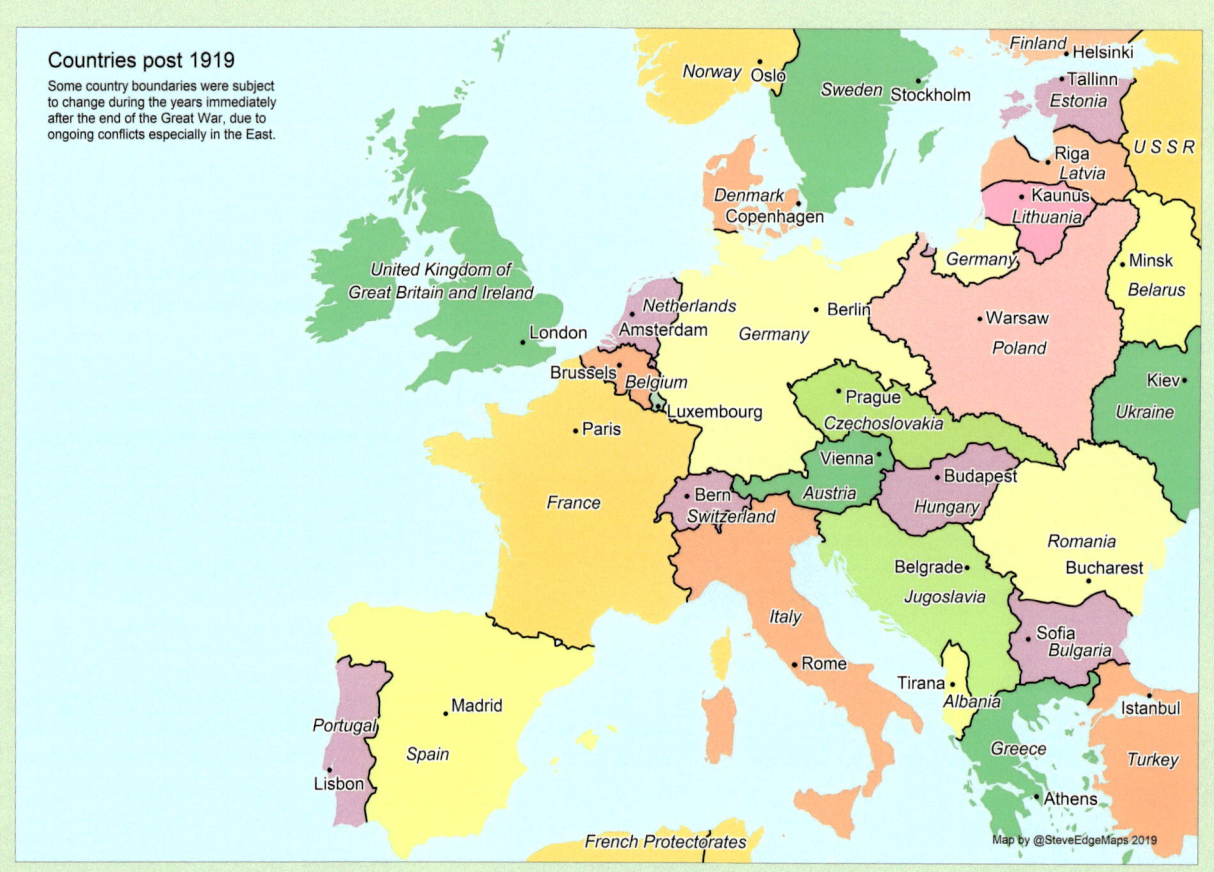

Map 2: Changes in European country boundaries after 1945 and 1994

After the Second World War further German territory was handed over to Poland, while an equivalent area of eastern Poland was annexed by the USSR. Otherwise the post-1919 boundaries were largely re-established, although the countries of Eastern Europe were dominated by the Communist bloc's 'Iron Curtain'. The end of Communism in the early 1990s led to the emergence of further new countries that had once formed part of either the USSR, Czechoslovakia or Jugoslavia, as shown here.

Many British and American war engines were used in the Middle East, including Persia (now Iran), Iraq, Palestine (much of it now in Israel) and Egypt. As their activities here are not central to the contents of this book no map is included. Most readers will be familiar with the many conflicts that have affected these countries since 1945.

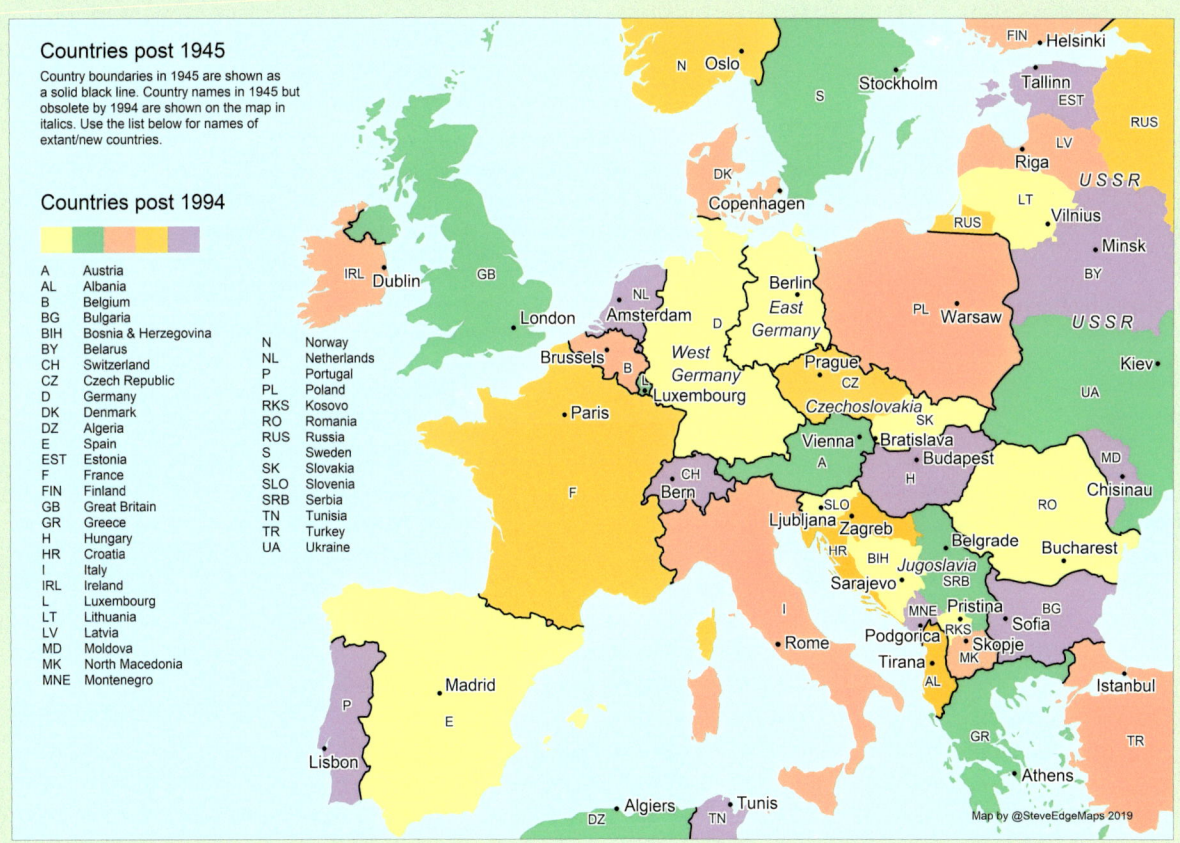

General introduction

To describe all of the classes of steam engine involved in the Second World War would require a book running to many volumes and, as listed in the Bibliography, several such publications already exist. This book describes the 12 classes, five German, five British and two American, produced in the greatest numbers by the main protagonists. In addition, other lesser German War Engines are discussed in Chapter 2.3.

The period of history occupied by the Second World War differs depending in which country you happened to be living at the time. For us in Great Britain, and in our then Empire, it started in September 1939 when Poland was invaded by both Germany and the USSR. In Austria it effectively began with its annexation by Germany in March 1938 (the 'Anschluss'), while Czechoslovakia was invaded in March 1939. America became involved in December 1941. The war in Europe ended with Germany's surrender in May 1945 (VE Day, 8 May 1945). For the purposes of this book any classes of German Deutsche Reichsbahn Gesellschaft (DRG) steam engines built from 1938 are considered as part of Germany's war effort, although the true 'Kriegsloks' did not appear until 1942. The British and American 'Austerity' War Engines also started to appear in that year.

German 'Kriegsloks' continued to be built after 1945 by countries outside Germany, while both types of 0-6-0 tank used by the War Department (WD) and United States Army Transportation Corps (USATC) were built for industrial use into the 1960s. All of the engines that feature in this book contributed greatly to the railways of Great Britain and Europe, many until the end of steam in the relevant country. Their influence can still be seen today on heritage and tourist railways throughout the Continent. The 12 main classes should also be well known to those British railway enthusiasts who, like me, after 'trainspotting' in the 1950s and '60s, extended their interest in steam engines to the Continent after 1968.

My interest in these engines also stemmed from my inheriting from an uncle a complete collection of *The Railway Magazine* published during the war. Despite the obvious difficulties, the magazine continued to be published throughout the conflict and included detailed accounts of the new war engines introduced in Britain and the USA during this period. Remarkably, one issue also gave detailed specifications for the chief German engine, the Class 52 'Kriegslok', almost 7,000 of which were built. With the kind permission of Mortons Media Group these reports are reproduced in the relevant chapters.

The DRG, WD and USATC engines described were also used in both the European and Asian parts of Turkey. For completeness these have also been included, while their deployment elsewhere in the Middle East is mentioned only briefly. It is worth noting that, being neutral during the war, Turkey received new engines from both sides of the conflict! Details of the careers of these engines after the war are given in the relevant chapter.

The methods used to number these engines varied from country to country, and the differences are summarised in Appendix 1. Appendix 2 lists the many abbreviations used in the text while Appendix 3 provides a Bibliography.

Tables 1 to 5 provide selective lists of the large number of engines that have been preserved. Tables 1 and 2 include the many preserved ex-DRG engines, many as part of large museum collections. These are therefore listed by preservation site, while other engines, shown in Tables 3 to 5, are listed by class. It should also be borne in mind that throughout the period parts, including boilers, were exchanged between engines. This applies particularly to those 'Austerity' saddle tanks that worked for the British National Coal Board after the war. Some preserved engines also frequently move from site to site. The locations given in the tables are therefore based on the best information available in 2020.

Part 1 Kriegsdampflokomotiven – German War Engines

Introduction

By 1938 a series of standard steam engines were being built in numbers for the German State Railway, Deutsche Reichsbahn Gesellschaft (DRG), to the design of their CME, Dr Richard Wagner. After the Nazi invasion of the USSR (Operation 'Barbarossa') in June 1941 a critical shortage of motive power became apparent. As a result Hitler demanded the construction of 15,000 new steam locomotives to be built using the minimum of materials and labour within a year. The result was a race to develop a series of 'Austerity' engines. Initially the DRG's Class 86 2-8-2 tanks and Class 44 and two-cylinder Class 50 2-10-0s were selected as 'Transitional War Locomotives', or 'Übergangskriegslokomotiven' (ÜK). Non-essential features such as pre-heaters, feed pumps and smoke deflectors were omitted, while the usual spoked leading wheels were replaced by disc ones. The next stage was even more radically simplified 'Kriegsdampflokomotiven' (KDL), the 'War Steam Locomotives'. The most important of these was the 'Austerity' version of the Class 50, the Class 52 'Kriegsloks', classified 'KDL 1' (see Chapter 2.1). In all, 13 KDL classes were identified, all of which are included in Chapter 2.3 (motorised and electric locomotives were classified KML and KEL respectively). The German engines retained their DRG class and were numbered in sequence (see Appendix 1).

By the end of the war some 12,700 2-10-0s alone had been produced. This was accomplished through the commandeering of the railway works of occupied countries so that more 'Kriegsloks' could be built, often using slave labour. Poland was particularly affected, but German war engines were also built in Austria, Czechoslovakia, France and Belgium.

The end of the war
The end of the war in 1945 found these engines scattered across Europe, many of them damaged. As the invading armies closed in on Germany and more and more countries were liberated, their railways were initially administered by the occupying army. In the West, control was in the hands of the British, American and French armies. Once civilian rule had been restored many captured German locomotives were slowly returned to the newly established Deutsche Bundesbahn (DB). Both France and Belgium, however, retained as war reparations most of those that had been built in their factories.

In the Eastern Sector occupied by the USSR, Germany and its allies were systematically stripped of as much of their railway equipment as possible. This was organised by the USSR's Ministry of Transportation (Ministrvo Putej Schrobtenia (MPS)). Thousands of steam engines were taken to the USSR and designated 'Te', where the 'T' stands for 'Trofyeinyi' or 'booty'. As described later, many were returned to the USSR's Eastern bloc allies during the Cold War. The Polish PKP had a large stock of German engines due to the country being shifted bodily to the west into German territory. Here the railways were taken over by the PKP (the USSR took over 20% of Polish territory in the east).

The post-war period and preservation
In 1949 Germany was officially split into two: West Germany (occupied by France, the USA and Great Britain) and East Germany (occupied by the USSR). The two halves were later to be separated by the 'Iron Curtain'. In the West Deutsche Reichsbahn Gesellschaft (DRG) was renamed Deutsche Bundesbahn (DB), while in the East the title Deutsche Reichsbahn (DR) was retained. Both DB and DR converted their ÜK engines back to their original designs, as did the other countries where they worked.

During the 1950s DR rebuilt many of its steam locomotives to prolong their lives, while DB concentrated on electrification or dieselisation. Steam therefore lasted much longer in East Germany, as it did in other Eastern European countries. Even before the end of Communism western enthusiasts were purchasing these engines for preservation. This became a flood when Communism ended in 1990. In many ways Eastern Europe can be equated with Barry scrapyard as a source of motive power for the Continent's heritage railways. In addition, many of the former Eastern bloc countries realised the tourist potential of running steam specials and restored some of their steam engines for this purpose.

DB and DR were combined to form a new organisation, Deutsche Bahn, in 1994.

1 The German 'Übergangskriegslokomotiven' or Transitional War Locomotives (ÜK)

1.1 The DRG Class 86 2-8-2 tank engines

These well-proportioned 2-8-2 tanks were used on freight and semi-fast passenger trains on both main and branch lines, and 775 were built for DRG between 1928 and 1943. They weighed 88.5 tonnes in working order, had an axle-load of 15.6 tonnes and a top speed of 70km/h. They were built by most of the German manufacturers together with, from 1939, WLF in Austria. Some 166 simplified ÜK versions were built from 1942. The largest number (73) was built in occupied Poland by the Deutsche Waffen und Munitionsfabriken (DWM), formerly the Polish firm of Cegielski, Poznan. This factory produced the first and last ÜK engines numerically: Nos 86.456 in 1942 and 86.875 in 1943. ÜK engines were also built by Krupp, BLW and BMAG in Germany. Other German factories continued to build the standard engines during 1942, as well as WLF in Austria.

Post-war distribution

It appears that 76 engines went, via the MPS, as war trophies to the USSR.

Distribution was otherwise as follows: 380 in West Germany (DB Class 86), 163 in East Germany (DR Class 86), 27 in Austria (ÖBB Class 86), 46 in Poland (PKP Class TKt3) and 46 in Czechoslovakia (CSD Class 455.2). All were withdrawn by the mid-1970s, although some later found use as stationary boilers or industrial engines.

To increase the production of steam engines at the start of the war, Germany introduced the simplified 'Transitional War Locomotives' (ÜK). These were the precursors of the true 'Austerity' 'War Locomotives' (KDL) that appeared from 1942. The three DRG classes involved were the two-cylinder Class 86 2-8-2 tank, the three-cylinder Class 44 and two-cylinder Class 50 2-10-0s. Some 770 Class 86 2-8-2 tanks were built between 1928 and 1942, with later engines built as ÜK locomotives. One of the wartime engines, but built in Austria before the introduction of the ÜK machines, No 086.400.9 (ex-DRG No 86.400 (WLF No 9243, 1941)) was one of 380 engines that went to the West German DB after the war. It is seen here outside its shed, possibly at Schweinfurt, in July 1972. The engine was withdrawn in August 1973.

Another Austrian-built Class 86 2-8-2 tank (ex-DRG No 86.781 (WLF No 9501, 1942)) was one of 27 engines that remained in Austria after the war. Here, running as ÖBB No 86.781, it heads downhill towards Hieflau with a train from Eisenerz on 5 September 1969. The masts for the line's electrification are already in place. No 86.781 was withdrawn in March 1972.

A number of the Class 86 engines were built in Poland. One example, ÖBB No 86.476 (ex-DRG No 86.476 (DWM/Cegielski No 461, 1942)), stands outside Hieflau station in Austria after arrival with a train from Eisenerz on 5 September 1969. Originally built as a ÜK loco, No 86.476 was also withdrawn in March 1972. It is now preserved, but unrestored, at the ÖGEG Railway Museum at Ampflwang. Forty-six engines remained in Poland after the war, classified PKP Class TKt 3.

Preservation

Wikipedia's 'List of Preserved Steam Locomotives in Germany' shows that nine Class 86 engines are preserved in Germany, two of which are ex-ÜK engines. One of these, No 86.457 (DWM No 442, 1942) was preserved by DB prior to German reunification and is now part of the German National Collection. In 2005 it was damaged during a fire at the Nuremberg Railway Museum and is now at Heilbronn for cosmetic restoration.

The second ex-ÜK engine, owned by the Dresden Transport Museum, is ex-DRG No 86.607 (BLW No 15280, 1942), which is now in the care of the Vogtland Railway Association at Adorf, Saxony. Another, ex-DRG No 86.001 (MBK No 2356, 1928), was the first of its class and was selected by the DR as one of its active museum fleet ('Traditionsloks'). These was taken over by DB after reunification (see Table 1) Three further examples, including one ex-ÜK engine, are preserved in Austria, while a 1935 member of the class, latterly PKP TKt3.16, is preserved in Poland (see Table 2).

1.2 The DRG Class 44 2-10-0s

During the Second World War these heavy three-cylinder 2-10-0s were used on heavy freight on main lines throughout Germany and its occupied territories. In all just under 2,000 engines were built for DRG between 1926 and 1949. The locos alone weighed 95.9 tonnes in working order, with an axle-load of 19.3 tonnes and a top speed of 80km/h. Weight precluded them from many of the German main lines of the time, so initially only ten were built. Construction began again in 1937 and more than 700 were built by German manufacturers before the first ÜK versions appeared in 1942. The first of these numerically was No 44.728, built in France by Schneider (Works No 4673). Other French manufacturers that produced ÜK engines during 1942 included Fives at Lille, Batignolles in Paris, and SFCM at Mulhouse. Many were also produced by German factories together with WLF in Austria.

In all some 2000 engines were built up to No 44.2025 (Fives, Lille No 5143) in 1946. Other French factories were involved in building these engines at this time and many were retained by SNCF as war reparations after the war. The last engine to reach Germany was No 44.1858, also built by Fives at Lille (No 5062), in 1944. Ten engines (Nos 44.1231 to 44.1240) had earlier been ordered from the Danish firm of Frichs, based at Horsens. These were never finished and in December 1944 they were transferred to BLW in Berlin, although they were not delivered to the DR until 1949. By this time BLW had been nationalised by the

Right: One of the wartime Class 44s, now DB No 44.340-8, is seen on the turntable at Rottweil depot on 26 November 1971 during a visit organised by Bill Alborough's TEFS. It was built as a ÜK engine in 1941 (ex-DRG No 44.1555(BLW No 15394)). It was renumbered in 1960 and withdrawn in August 1973..

GDR to become LEW.
Some of the class were rebuilt for oil-burning on both DB and DR.

Post-war distribution

It appears that only six engines were taken to the USSR by the MPS as 'war trophies', others were passed by the Soviet to their allies. Distribution was approximately as follows: 1,242 in West Germany (DB Class 44), 335 in East Germany (DR Class 44), 132 in Poland (PKP Class Ty4), 4 in Romania (CFR Class 150) all ex-ÖBB, 13 in France (SNCF Class 150X) plus 226 still under construction there in 1945, most of which were scrapped. The Class was not popular in France and, in 1955, the SNCF sold 48 engines to Turkey (TCDD Class 56701). The engines worked into the 1950s in France, the 1960s in Romania, the 1970s in West Germany and the 1980s in Poland, East

Right: DB No 44.943-9 heads a heavy eastbound freight for Kassel through Soest, a city in North Rhine-Westphalia, on a wet 30 September 1969. This ex-DRG engine, No 44.943, was built in France as a ÜK loco by SFCM, No 4292, in 1942. It was withdrawn from Gelsenkirchen-Bismarck depot in February 1972. *Barry Payne*

Germany and Turkey. The DR rebuilt many of their Class 44s as oil-burners in 1968 although these later reverted to coal burning. Some of the engines later found use as stationary boilers or industrial engines.

Preservation
Wikipedia's 'List of Preserved Steam Locomotives in Germany' shows that 43 Class 44 engines are preserved in Germany. These include 19 ex-ÜK engines including ex-DR No 44.1093 (WLF ÜK No 9449, 1942). This engine was once part of the DR's active museum fleet (Traditionsloks), and is now part of the German National Collection (see Table 1). Elsewhere ex-ÜK examples are preserved in Austria, the Netherlands and Turkey (see Table 2).

Above: DB Class 44 2-10-0 No 44.380-4 heads a freight through Cochem in the Mosel Valley towards Trier on 25 September 1972. The engine (DRG No 44.1380) was built as a ÜK loco by Krupp in 1942 as No 2802. It was withdrawn from Ehrang depot in August 1973. *Barry Payne*

Left: Another DB Class 44, No 44.250-9, heads a southbound freight through Cochem in the Mosel Valley on 25 September 1972. The engine (DRG No 44.1250) was also built as a ÜK loco by BLW in 1942, as No 15236. It was withdrawn from Koblenz-Mosel depot in March 1974. Note the vineyards on the hillside behind the train. *Barry Payne*

DRG Classes 86 and 44 in preservation

Above: Two ex-DRG engines originally built as ÜK locos are seen working together at Tübingen during the celebrations of 25 years of steam preservation on the Zollernbahn on 13 April 1998. The first engine, ex-DRG Class 44 2-10-0 No 44.1616 (Krenau/Chrzanow No 1104, 1942) was withdrawn in 1991, after use as a steam-heating unit. It was then purchased for preservation on the Zollernbahn but has since moved to the South German Railway Museum at Heilbronn. The second engine is another Polish-built example, ex-DRG Class 86 No 86.457 (DWM/Cegielski No 442, 1942), which later became DB No 86.457 (No 86.457-9 after computerisation). On withdrawal in July 1972 it went to the German Railway Museum at Nuremberg. After spending time at the South German Railway Museum at Heilbronn it returned to Nuremberg, where it was damaged by the fire of October 2005. It is now back at Heilbronn for restoration and repainting.

Above: Two ex-DR Class 44s, double-heading a 2,000-tonne gravel train from Immelborn to Eisenach, are seen near Ettenhausen in April 2014 during a Plandampf in the area. A further engine is banking in the rear. The leading engine is No 44.1486-8 (ex-DRG No 44.1486 (Schneider No 4728, 1943)), originally an ÜK loco. After German reunification, it was restored as a DB museum engine until finally withdrawn in 1994. It is now preserved by the Eisenbahnfreunde Traditionsbetriebswerk based in Stassfurt. *David Alison*

1.3 The DRG Class 50 2-10-0s

The third and most important class of simplified 'Transitional War Locomotives' (ÜKs) were the two-cylinder DRG Class 50 2-10-0s. The first, No 50.001, had emerged from the Henschel Works as No 34355 in Kassel in March 1939. The locos weighed 86.9 tonnes in working order, with an axle-load of 15.2 tonnes and a top speed of 80km/h. Their lower axle-load made them suitable for both main-line and branch-line work. More than 3,140 were built between 1939 and 1943, initially with full 'Wagner'-type smoke deflectors. Many were built in the works of the occupied countries including Austria, Belgium, Czechoslovakia, France and Poland. The first of the ÜK engines, No 50.1583 (Henschel No 26393), appeared in March 1942. Initially only a few features were simplified, while the smoke deflectors were dispensed with. However, to speed up production further, more and more modifications were made until the engines resembled the ultimate 'Kriegsdampflokomotiven' (KDL 1), the Class 52s, which succeeded them from October 1942 (an example is CFR No 150.1105, see Page 44). However, both Krupp and MBA continued to build them into 1943. In addition, between 1944 and 1948 24 ÜK engines were built by Belgian factories, some of which were used by SNCB as its Class 25 (see below).

A remarkable collection of ex-DR 2-10-0s and other main-line engines has been gathered by Herr Bernd Falz at the former engine shed at Hermeskeil. These include nine ex-DR Class 44 2-10-0s. Only one, No 44.1106-2 (ex-DRG No 44.1106 (BLW No 15155, 1942)), has been fully restored. Still languishing in its unrestored state is ex-DR No 44.1537-9 (ex-DRG No 44.1537 (BLW No 15376, 1942)), seen at Hermeskeil on 13 September 2014. Both were originally ÜK locos.

Above and below: After the end of the war, when Germany was divided into East and West by the Iron Curtain, many Class 50s remained in both countries. In the West, despite increasing dieselisation and electrification, they continued to work for DB until 1977. Here Class 50 2-10-0 No 053.101-2 (ex-DRG No 50.3101 (MBA No 14227, 1943)), originally built as a ÜK loco, is seen at Plattling in Bavaria with a passenger train from Landshut on 13 July 1972. The engine was withdrawn in July 1975.

The Steam Engines of World War II in Europe

Steam was also active around Rottweil in Baden-Württemberg during the early 1970s. No 052.891-9 (ex-DRG No 50.2891 (BMAG No 11947, 1942) stands at Rottweil station with a freight for Stuttgart on 27 November 1971. Originally built as a ÜK loco, it was withdrawn from Rottweil depot in September 1974. Note that there is a built-in guard's compartment in the tender. The freight included a BR ferry van that advertised 'Through to the Continent with British Rail'.

DB No 052.953-7 is seen on Tübingen shed on 23 September 1972. The ex-DRG engine, No 50.2953, was built as a ÜK loco in Austria by WLF as No 9540 in 1942. It was withdrawn in June 1976. *Barry Payne*

Post-war distribution

Some 60 engines were taken by the MPS for use in the USSR, although some of them were later passed to its Eastern European allies, including Jugoslavia, Romania and Poland. A large number of other engines were badly damaged during the war were written off or used as spares to produce one working loco. These included a large number of French-built engines that were returned to France as war reparations.

Distribution was otherwise approximately as follows: 2,160 in West Germany (DB Class 50), 340 in East Germany (DR Class 50), 13 in Austria (ÖBB Class 50), 23 in Belgium (SNCB Class 25), nine of which returned to the DB in 1950, 50 in Bulgaria (BDZ Class 14) including 20 from the CSD in 1959/60, 28 in Czechoslovakia (CSD Class 555), prior to 1959/60, 36 in France (SNCF Class 150Z), many of which later returned to DB, 3 in Jugoslavia (JDZ Class 33), 58 in Poland (PKP Class Ty 5) and 18 in Romania (PKP Class 150), many from the MPS, Hungary and Austria.

Modifications, rebuilds and new builds

After 1950 both DB and DR rebuilt or otherwise modified at least some of their Class 50s. In 1955 31 were rebuilt by Henschel for DB with Franco-Crosti boilers and renumbered 50.4001 to 50.4031. (In the same year BR fitted ten of its 9F 2-10-0s, Nos 92020 to 92029, with the same boiler, although these were later sealed off.) Two, Nos 50.4001 and 50.4011, were converted to oil-burning. The next significant modification was in 1961 when compartments to accommodate freight guards were built into 730 of their tenders (Kabinentender). In addition, both

DB and DR replaced the full 'Wagner'-type smoke deflectors with the smaller 'Wittemodel' as fitted to many of BR's Class A3 'Pacifics', including No 60103 *Flying Scotsman*, during the 1960s.

In 1956 it was decided that DR needed still more goods engines, so 88 more were built to an improved design. To save costs many parts were interchangeable with DR's Class 23.10 2-6-2, first introduced in 1955. Plate frames were used together with a newly designed welded steel boiler. The engines were classified 50.40 (not to be confused with the DB Class 50.40s fitted with Franco-Crosti boilers). The engines were built by the Karl Marx Loco Works (KLM) in Berlin as Nos 50.4001 to 50.4088. When No 50.4088 was delivered in 1960 it became the last standard-gauge steam engine built in Germany, although the weakness of the locomotives' plate frames led to their early withdrawal.

DR's rebuild of many of its original Class 50s was more successful. Between 1959 and 1963 208 were fitted with the new welded boilers and reclassified Class 50.35. In many cases a Giesl ejector was also added. The first engine to be so treated, by KLM, was No 50.380 (Borsig No 14970, 1940), which became No 50.3501. In addition, between 1966 and 1971 72 of the engines were converted to oil-firing and reclassified Class 50.50, numbered 50 0001 to 50 0072. An increase in the cost of crude oil, however, led to their withdrawal in 1981. Finally, between 1946 and 1960 Romania built 282 engines based on the DR Class 50s for CFR (Class 150). Later engines were fitted with a double chimney and Kylchap exhaust. Other engines were built in Romania for China and North Korea.

The engines worked into the 1950s in France, the 1970s in West Germany and Austria, and the 1980s in Romania, Poland, East Germany and Turkey. Some later found use as stationary boilers or industrial engines, including three Austrian engines that were

DB No 53.010-5 arrives at Lichtenfels with a train from Horb on 21 September 1972. This ex-DRG engine, No 50.3010, was built as a ÜK loco by BLW, No 9597, in 1942. It was withdrawn from Hof depot in September 1974. *Barry Payne*

sold to the Graz-Köflacher Bahn (GKB) in 1972. Many were retained in Bulgaria and Romania as a strategic reserve, although almost all have now been broken up.

Preservation
A large number of Class 50s are preserved, many as static museum locos, while some, particularly the Class 50.35s, are at work on heritage railways across Europe. The Wikipedia website 'List of Preserved Steam Locomotives in Germany' lists more than 20 original and 44 rebuilt Class 52.35s in Germany alone. Nine of the original engines and nine of the rebuilds were built as ÜK engines. Both of the DB or DR museum engines, No 50.622 and DR No 50.(1)849, were built before the introduction of ÜK engines.

One of the DR's 1960s-built Class 50.40s, No 50.4073, has also been preserved together with one of its oil-burning Class 50.35s, No 50.0072 (see Table 1).

The only engines in original condition preserved

outside Germany are in Austria (No 50.1171, the last engine produced before the Skoda factory in Czechoslovakia started to turn out ÜK engines), Bulgaria, Poland and Turkey. Other examples in Austria, Belgium, France and the Netherlands are rebuilt Class 50.35s. These include No 50.3666 (as oil-burning No 50.0073), originally built as a ÜK engine (ex-DRG No 50.2145 (Franco-Belge No 2567, 1943)). Several post-war Romanian-built examples are also preserved (see Table 2).

DRG Class 50s officially preserved in Germany by DB and DR pre-1990

Both DB and DR preserved Class 50s as part of their 'Heritage' fleets. Neither were ÜK engines. DB's museum engine, No 50.622, has arrived at Tübingen at the rear of a special from Stuttgart via Horb on Saturday 11 April 1998. The engine was built as DRG No 50.133 (BLW No 14862, 1940). In September it went to the Belgiam SNCB as their No 25.014, but returned to the DB in May 1950 where it was, for reasons unknown, renumbered 50.622. The original No 50.622 became No 50. 2847. The train was bringing enthusiasts to the celebrations of 25 years of steam preservation on the Zollernbahn. Housed at the Nuremberg Railway Museum, the engine was badly damaged during the fire that engulfed it in October 2005.

Between 1959 and 1963 DR rebuilt 208 of its Class 50s as Class 50.35 (see photos below). One of the original engines, No 50.1849-4 (ex-No 50.849, Krauss Maffei, Munich No 16058, 1940), was retained as a 'Traditionslok' for special workings. It is seen here leaving Berga with a freight from Gera to Weischlitz during the 'Vogtland Plandampf' on 10 October 1993. In the distance the steam-hauled 15.16 from Weischlitz to Gera is seen departing the station behind DR 2-8-4T No 65.1049 (LKM No 121049, 1956). Both were then operating as DB Museum Locos and are now part of the German National Collection. No 50.1849-4 is currently based at Glauchau in Saxony, fitted with large 'Wagner' smoke deflectors typical of the original DRG Class 50s. No 65.1049 is housed at the Nuremberg Railway Museum

Class 50.35s at work in East Germany for DR

One of DR's rebuilt Class 50.35s, No 50.3559-7 (ex-DRG No 50.1486 (Henschel No 26296, 1941)), is seen passing Hadmersleben with a very mixed freight from Magdeburg to Halberstadt in July 1988. The 208 Class 50 engines were rebuilt between 1959 and 1963 and fitted with new welded boilers. This engine worked the DR's very last scheduled steam-hauled passenger train from Halberstadt to Thale in October 1988. The train was normally diesel-hauled, but was turned over to steam for the occasion. After withdrawal in October 1991 the engine was purchased for use on the 'Villa Express' at Erftstadt-Liblar, North Rhein Westphalia, although this enterprise may have since closed. In 2016 it was reported as under restoration at the MaLoWa works at Benndorf in Saxony-Anhalt. *David Alison*

DR Class 50.35 No 50.3662-9 (ex-DRG No 50.1249 (WLF No 9183, 1941)) is pictured at Gunsleben ready to return with a passenger train to Oschersleben. Gunsleben was once a wayside station between Oschersleben and Jerxheim in Lower Saxony, but the Iron Curtain between West and East Germany split the line in two, with Gunsleben becoming the GDR terminus. At the time of David Alison's visit in July 1988 there were just three return workings per day (mostly mixed trains). These were the last regular standard-gauge steam-hauled scheduled passenger trains on DR. Sadly the line has since closed. After withdrawal in December 1991 the engine was purchased for the private railway museum at Hermeskeil. *David Alison*

Class 50.35s preserved in Germany

Left: Many ex-DRG engines that survived long enough to be preserved were used as steam-heating units after withdrawal. An example was ex-DR No 50.3522-5 (ex-DRG No 50.1368 (BLW No 15083, 1941)), which was converted to a non-mobile heating unit early in 1989 based at Pasewalk Depot until withdrawal a few years later. The engine was purchased for private preservation in 1993, and here it makes a rather sad sight at Röbel in north Germany, still unrestored, on 12 September 2010. The engine had been paired with a much older six-wheeled tender.

Below left: One of the preserved Class 50.35s seen in operation during the Dresden Steam Festival in the former East Germany was No 50.3648-8 (ex-DRG No 50.967 (Krupp No 2332, 1941)). Seen here, it is working a special train from Dresden to Neustadt (Sachsen) and crossing a viaduct near the village of Stolpen on Sunday 30 April 2000. The engine was withdrawn in January 1991 and is now based at the former Chemnitz-Hilbersdorf engine shed, run by the Saxon Railway Society.

Class 50.35s preserved in France and Belgium

Right: Ex-DR Class 50.35 No 50.3661-1 (ex-DRG No 50.1224 (WLF No 9158, 1941)) is a long-term restoration project for the Tourist Railway of the Haut Quercy at Martel in the Dordogne. It was withdrawn in August 1994 and eventually arrived at Martel in 2007. It is seen there on Sunday 19 April 2015.

Below: Other Class 50.35s were exported to both Belgium and the Netherlands. One of these was No 50.3666-0, originally built as a ÜK loco (ex-DRG No 50.2145 (Franco-Belge No 2567, 1943)). For a time it was fitted with a Giesl ejector and was withdrawn in May 1992. On 6 July 1996 it was working on the Vennbahn heritage railway on the Belgian/German border when it was seen crossing the main road (Route 68) at Malmedy with a train for Trois Ponts. The engine is now based at Apeldoorn in the Netherlands for use on the Veluwsche Steam Train. Here it has been restored as oil-burning DR No 50.0073-2.

Class 50s working and preserved elsewhere in Europe

Right: Seventeen Class 50s remained in Austria after the war, where they retained their original large 'Wagner' smoke deflectors. One of these, Czech-built ÖBB No 50.1171 (ex-DRG No 50.1171 (Skoda No 1250, 1941)) is seen crossing a minor road south of Linz on 2 September 1969. After completion of DRG No 1171, production at the Skoda factory switched to ÜK locos. The engine went to the Graz-Köflacher Bahn (GKB) in August 1973 and is now at the Brenner & Brenner Steam Locomotive Operating Co, where it is retained for special workings.

Right: One of the 54 Class 50s that ended up in Poland after the war was PKP No Ty5-10 (Ex-DRG No 50.451 (Schichau No 3413, 1940)). It is seen at Wolsztyn on Monday 1 May 2000, prior to its cosmetic restoration. Note that this engine also still carries its original 'Wagner' smoke deflectors.

Right: After the war the DRG Class 50 became one of Romania's standard classes (CFR Class 150). In all, 282 of the Class 150s were built at the Resita and the 23rd August Works (formerly the Malaxa Works) in Bucharest between 1946 and 1960. When this photo was taken at Subcetate on 20 June 1978, many had recently been retired and placed into 'strategic reserves'. The 'nameplates' carried on their 'Wagner'-type smoke deflectors read 'Resita', the place where 251 of them were built. These 'strategic reserves' remained in Romania until the late 1990s, but almost all of the engines have since been broken up, sometimes illegally!

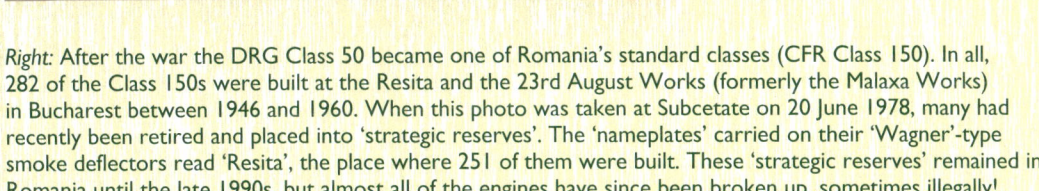

Above: In the 1990s a number of CFR's Class 150s were preserved in Romania. One of the engines, No 150.025 (Resita No 710, 1950), was restored to work special trains. Here the engine has been splendidly turned out to work a special chartered by the Rom Steam-Aldo organisation on behalf of the LCGB and WRS. from Brasov to Intorsura Buzaului on 10 October 1997. Romania's steam engines burned a mixture of coal and oil, hence the black smoke; the oil tank can be seen at the back of the tender. Economic problems in Romania have put the future of its preserved steam engines in doubt. No 150.025 is now in Bucharest.

Right: During the journey with the RCTS/WRS special, No 150.025 paused at the small station at Budila. Here harvested sugar beet was being brought to the station to be loaded into railway wagons. Although a few tractors were present, most had been brought to the station in wagons hauled either by horses or, as here, oxen. Some of the horses were getting a free lunch by eating the loose sugar beet!

2 Kriegsdampflokomotiven – War Steam Locomotives (KDL)

2.1 The DRG Class 52 'Austerity' 2-10-0s (KDL 1)

The Class 52 'Austerities' were to become by far and away the most famous and numerous of the 'Kriegsdampflokomotiven'. The first, DRG No 52.001 (BLW/Borsig No 15446), appeared in September 1942. The German Reich gave considerable publicity to its new engines in a press release of 13 October 1942. This explained that the rationale behind the design was to significantly reduce both the materials and manpower needed for their construction. Full details of their leading dimensions were also given. Although a state of war existed between Britain and Germany, a year later *The Railway Magazine* published the press release reproduced here.

Two types of 'Austerity' tenders were produced for the engines: the 'Wannen' and 'Steifrahmen' types. The former had a rounded 'bath tub' tank while the latter had a rigid frame and straight side tanks. In 1943 Henschel produced 137 'condensing' engines, numbered from 52.1850, which allowed the exhaust to be recovered in the smokebox and condensed in a large tender, which were fitted with both three-and two-axled bogies. As a result they could travel for 1,000km without taking water, although turning them was a problem! Forty-one further engines were therefore built, from No 50.1987, with smaller four-axled tenders.

Below: 'German "Austerity" Locomotive' – an extract from *The Railway Magazine* for September and October 1943. By permission of Mortons Media Group

German "Austerity" Locomotive

German "Series 52" austerity locomotive

CONSIDERABLE publicity was given in the German Press on October 13, 1942, to the new German war locomotive (*Kriegslokomotive*) which has been produced as a result of an Order made by Hitler in March of the same year. He placed the production of locomotives and rolling stock under the control of the Minister for Armaments and Munitions, and an immediate step was the development of a "transition" locomotive for production until the end of 1942. This was a simplified version of the "Series 50" goods locomotive, arranged to be specially suitable for use in the low temperatures encountered in winter on the Russian Front. An illustration of the "transition" locomotive is not available but the new *Kriegslokomotive* is shown in the picture. The new locomotive is being mass-produced throughout Germany and German controlled countries, and is classed as "Series 52."

The "Series 52" engines were in the first place designed for use in the severe climates of occupied Eastern Europe; the driver's cab is entirely closed and the gangway to the tender is provided with a concertina passage. Crews were expected to have to work the engine in long shifts, and consequently armchairs have been provided for both driver and fireman. A hammock is provided in the driver's cab for one man to rest while the other attends the fire.

The first of the completed locomotives together with a train of some new type goods wagons was sent on a tour to all the workshops engaged on building these engines. The route taken was from Berlin to Poznan, Elbing, Warsaw, Krakow, Vienna, Pilsen, Munich, Stuttgart, Cologne, Liége, Lille, Brussels, Duisburg, Kassel, Hanover, and back to Berlin.

The standard "Series 50" goods locomotive of the German State Railway was the most recently introduced type before the outbreak of war, and details were included for the first time in the 1939 edition of the handbook issued for the use of the Reichsbahn staff. This engine, which has the 2-10-0 wheel arrangement, is a superheated two-cylinder simple locomotive, of which the leading dimensions are as follow :—

Cylinders, dia.	600 mm. (23⅝ in.)
" stroke	660 mm. (26 in.)
Dia. of driving wheels	1,400 mm. (4 ft. 7⅛ in.)
" leading wheels	850 mm. (2 ft. 9½ in.)
Total wheelbase, including tender	18,890 mm. (62 ft.)
Boiler pressure	16 kg. per sq. cm. (227 lb. per sq. in.)
Grate area	3.9 sq. mm. (43 sq. ft.)
Total heating surface	177.6 sq. metres (1,910 sq. ft.)
Heating surface, superheater	63.6 metres (684 sq. ft.)
Weight in working order (engine only)	86,500 kg. (85 tons)
Adhesive weight	75,000 kg. (73¾ tons)†
Maximum permissible speed	80 km.p.h. (50 m.p.h.)
Dia. of tender wheels	1,000 mm. (3 ft. 3⅜ in.)
Water capacity	26 cubic metres (5,720 gal.)
Coal capacity	8 tonnes (7 tons 17 cwt.)
Weight of tender in working order	60,000 kg. (59 tons)

The plans of the "Series 52" engines had to make provision for the widest use of materials as are readily available in Germany, and for the most economical use possible of the reduced quota of steel; also they had to show a saving of labour in production. All precision work for these reasons has been cut down to a minimum, and the parts have been designed to allow mass production by sharing out a number of parts among many different factories. Copper has been replaced by steel; the firebox is entirely of steel. Coupling rods are made of rolled sections, with ends of pressed steel electrically welded to them. The weight on the driving wheels had to be kept as high as in the prototype; therefore economies had to be found in the selection of raw materials rather than in the amounts employed to make the finished parts of the locomotive; a plate frame instead of the original bar frame was adopted for this reason. The weight of the tender could be reduced to any desired limit; this has been achieved by making the tender sides serve for a frame, and by using two standard goods-wagon bogies. The tare weight has been brought down from 26 tons for the "Series 50" locomotive tender to 18 tons for the "austerity" tender, notwithstanding an increase in the coal capacity from 8 to 10 tons and in the water capacity from 26 to 34 cu. m. Smoke deflector plates on both sides of the smoke-box are missing, in addition to which there are no brackets for the headlamps, no sand box, and no bell. A number of hand rails and handles are also missing. There is, moreover, no feedwater heater, and many other devices not vitally important have also been omitted. The rodding, inclusive of the control and brake rigging, is simplified to the utmost. Of the usual 6,000 components employed in the "Series 50" locomotive, roughly 1,000 have been eliminated altogether, and 3,000 have been modified or altered. In comparison with the old construction, the new type represents a saving of 26 metric tons of materials, primarily steel and non-ferrous metals. Also, some 6,000 working-hours are saved. It is claimed that special constructional methods (which all locomotive works are now organising) enable a locomotive of the "war" type to be completed in half the time required for one of the original type. In view of the extensive manufacturing facilities available, the output is expected to be very high. The new locomotives are finished in unvarnished blue-grey, instead of in the familiar German livery of black with red wheels.

By December 1942 Henschel and BMAG/Schwartzkopff in Germany and WLF (Vienna) in Austria had already built 190. After this new engines started to appear from locomotive works throughout the areas of Nazi occupation. The number of works producing the engines in each country was as follows: Germany (9), Austria (1) Poland (2) and Czechoslovakia (1). During 1943 and 1944 many of the Class 52s were also sold new or sent on loan to Axis satellite countries, including Hungary, Romania, Croatia and Bulgaria. In addition they were also sent to France, Belgium, Serbia and Poland, which were occupied by German troops. Turkey, which was neutral, was loaned 53, which were later taken into TCDD stock as Class 56501.

At the end of the war construction had only just started in Belgium, where four works – Cockerill, Franco-Belge, Haine St Pierre and Tubize – produced 25 engines each between 1945 and 1951. In total some 6,700 Class 52 engines are thought to have been built.

Post-war distribution
At the end of the war Class 52s could be found throughout most of Europe. Initially, as the war ended, there were endless comings and goings of engines between territories held by both the Western Allies and the Soviets. As with the UK engines, many were taken by the MPS for use in the USSR; some 2,100 engines were involved, including 750 that remained in Soviet territory as the German Army withdrew, many of them damaged. Some of these were later cannibalised to make around 38 new engines. A further 500 were taken by the USSR as 'war trophies' from Hungary, Romania, Poland, Czechoslovakia and Austria. Most were converted to the USSR's 5-foot gauge. In 1946 the USSR returned to Poland many of the class built in its two factories, Cegielski/DWM and Chrzanow. These became PKP Class

Unrebuilt DR Class 52s retained in East Germany

While the West German DB withdrew its last Class 52s in 1963, in the East DR rebuilt 200 of them as Class 52.80. One of the few engines to remain in its original condition was No 52.6666 (Skoda No 1492, 1943). It was retired from Jüterbog depot in 1970, but was retained as a DR 'Traditionslok'. It is pictured working a 'German Model Railway Club' special on a circular route starting from Leipzig in July 1987. The engine is still in working order, based at the ex-DR depot at Schöneweide, Berlin.
David Alison

Another unrebuilt DR example, ex-DRG No 52.4924-8 (MBA No 13994, 1943), also became a 'Traditionslok'. It is seen at Cranzahl in Saxony during the celebrations of 100 years of the 750mm-gauge Cranzahl to Oberwiesenthal line on 19 July 1997. The engine had been withdrawn from active service in April 1990 and is now housed at the Sächsisches (Saxon) Railway Museum at Chemnitz (formerly Karl-Marx-Stadt).

Between 1953 and 1958 DR converted 25 of its Class 52s (as Class 52.90) to burn pulverised brown coal (or lignite), a fuel that was plentiful in the GDR. One of these, No 52.9900-3 (ex-No 52.4900 (MBA No 13970, 1943)), was withdrawn in 1975 and subsequently preserved by the Dresden Technical Museum. Its modified tender was photographed at Dresden's Altstadt depot during the annual 'Steam Festival' there on 30 April 2000.

Rebuilt DR Class 52.80s at work in East Germany

One of the DR's 200 rebuilt Class 52.80s, No 52.8144-9 (ex-No 52.1648, SACN 7915, 1943), passes Haldensleben with a freight in July 1985. At that time the engine was allocated to Haldensleben depot, but was soon transferred to Halberstadt, from where it was withdrawn in August 1987. *Adrian Cornish*

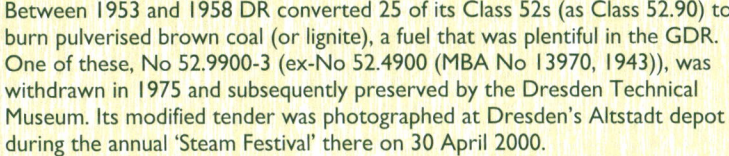

Ty42 to distinguish them from those built abroad (PKP Class Ty2). In 1947/48 it was the turn of the USSR's communist ally, Jugoslavia, to receive a batch of engines (JDZ Class 33). After the Warsaw Pact was established by the USSR in May 1955, more of the engines were returned to its members, including Hungary, Czechoslovakia, Poland and the GDR. The number of Class 52s available for work in the many countries is difficult to determine, particularly as the published information is often contradictory. Approximate details for each country are as follows:

West Germany (DB Class 52): The DB inherited some 900 Class 52s. Many of these were badly damaged but by 1951 50 had returned from Italy, along with four condensing engines from elsewhere. The class was not favoured by the DB and in 1951/52 40 were returned to France., where they had been built, while 36 were sold to Jugoslavia. Apart from a few which had been fitted with feedwater heaters by Henschel after the war all DB engines were withdrawn by 1954. In 1957, following a plebiscite in the disputed Saarland favouring its return to Germany, the DB acquired 14 more of the Class. Some of these lasted into the early 1960s, along with the water-heated engines.

East Germany (DR Class 52): There are thought to have been around 1300 Class 52s in East Germany in 1946 but, after some 485 were taken by the MPS and others were deemed unrepairable, the DR was left with just over 800 engines. Thirty of these went to Bulgaria in 1956, becoming BDZ Class 15. In 1962 the DR found itself short of locomotives so 60 more engines were purchased from the USSR. The class was withdrawn in the 1970s and 80s. As described below 200 were modified or rebuilt, some of which lasted to the end of DR steam.

The Steam Engines of World War II in Europe

Austria (ÖBB Classes 52 and 152): Austria was left with around 300 engines after some 450 were lost to the USSR as 'war trophies'. The ÖBB machines included 36 with bar-frames, which were reclassified Class 152; the 271 with plate-frames remained as Class 52. The Class 152s lasted longest, the last being withdrawn in the early 1970s. By then 13 Class 152s had been sold to the GKB and lasted until 1978.

Belgium (SNCB Class 26): The 100 engines built by Belgium's four locomotive works became SNCB Nos 26 001 to 100, although ten of these were later sold to Luxembourg. The last SNCB engines were withdrawn in 1963.

Bulgaria (BDZ Class 15): In total BDZ owned 275 of these engines, BDZ Nos 15.01 to 275. These had come from DR, CSD and the USSR. Some lasted until the end of steam in the 1980s.

Czechoslovakia (CSD Class 555): After the predation by the MPS, CSD was left with 180 engines. Some of these were converted to 5-foot gauge to work on cross-border freights to the USSR. In 1959 20 (some sources say only 18) were exported to Bulgaria and two to Jugoslavia. In 1962/63 100 more engines came from the USSR, where they had been converted from 5-foot gauge to standard gauge at CSD's expense. The last of the class was withdrawn in 1973.

France (SNCF Class 150Y): SNCF inherited around 40 working Class 150Ys, including some that remained in France after the war while others had been built in France by SACM. Ten of these engines were sold to Luxembourg in 1946/47. Around 40 French-built engines were also returned to France by the DB, but these were scrapped. The last of the class was withdrawn by SNCF in the late 1950s, although a few, sold for industrial use, survived until 1961.

Top left: Another rebuilt DR Class 52.80 crosses Putzkov Viaduct on the Bischofswerda to Bautzen line with a freight in October 1987. *Adrian Cornish*

Left: Although steam working officially ended on DR in 1988, many engines were retained as a strategic reserve. Some of these were used for special workings, together with DR's historic 'Traditionsloks'. Rebuilt DR Class 52.80 No 52.8200-9 (ex-DR No 52.467 (BLW No 15564, 1943)) stands at Zittau with an enthusiasts' special for Dresden. The engine, then still owned by DR, was withdrawn on 20 January 1993 and is now preserved at Mariembourg in Belgium (see page 28).

Right: In a smoky scene at Dresden's Alstadt depot on 29 April 2000, rebuilt DR No 52 8075-5 (ex-DRG No 52.1292 (DWM No 733, 1944)) reverses off the turntable during the annual Steam Festival. The engine was also withdrawn from Schöneweide depot, Berlin, in December 1993 and is now operated by Bahn-Nostalgie in Thuringia.

Right: Among the engines preserved at the former engine shed at Hermeskeil is rebuilt ex-DR No 52.8090 (ex-DRG No 52.7778 (MBA No 14362, 1944)), withdrawn from Cottbus depot in August 1990. Apart from its painted smokebox door, the engine is unrestored.

Ex-DR Class 52.80s preserved in Germany, Belgium and the Netherlands

Above: Since their withdrawal in the 1990s, many of the Class 52.80s have been preserved in Germany and elsewhere. Rebuilt DR Class 52 No 52.8087 (ex-DRG No 52.2455 (Henschel No 27623, 1943)) is seen at Hechingen after arrival with a special train from Tübingen on Sunday 12 April 1998 during the celebrations of 25 years of steam preservation on the Zollernbahn. After withdrawal from Schöneweide depot, Berlin, in December 1993, the engine was purchased by the Swabian Steam Society at Neuoffingen.

Right: Seen earlier at Zittau in 1991 (see page 26), Class 52.80 No 52.8200-9 was withdrawn from Bautzen depot in January 1993 and exported to Belgium. It is seen here passing Rolinvaux-Olloy between Mariembourg and Treines during a gala at the Chemin de Fer à Vapeur des Trois Vallées (Three Valleys Heritage Railway) on 22 September 1995.

Far right: One of the Class 52.80s, No 52.8160, is preserved in the Netherlands. The engine, ex-DRG No 52.532 (BMAG No 13099, 1943), went to the USSR after the war (becoming SZD No TE.532), but returned to the DR in 1962 and was later rebuilt. It was withdrawn from Zittau depot in November 1992 and is now preserved at the South Limburg Steam Train Company in the Netherlands. It is seen on 22 December 2015 partly dismantled after arrival at the museum, and is still under restoration. *Courtesy of Jordy Brouwers, South Limburg Steam Train Company*

Austrian ÖBB Class 152s – the last working Kriegsloks in Western Europe

Right: After the war more than 2,100 of DR's Class 52 Kriegsloks were taken by the Soviet military as 'war booty', of which around 350 were from Austria. Although many were later returned to the USSR's Warsaw Pact allies, this did not include Austria. The Austrian ÖBB nevertheless ended up with around 300 engines, which were active into the 1970s. Here two of the class, both fitted with Giesl ejectors, come off a northbound freight at Hieflau and head for the depot there on 5 September 1969.

Above: Giesl ejector-fitted ÖBB Class 52.1189 (DWM No 603, 1943) heads a southbound freight through the Enns Valley towards Hieflau at Weissenbach St Gallen on 5 September 1969. Included in the freight is an electric loco, which can be seen behind the Class 52 – electrification of the lines around Hieflau was imminent. No 52.1189 was withdrawn in November 1974.

Above: ÖBB Class 52 No 52.798 (Henschel No 28123, 1944) comes off the Hieflau avoiding line to head north up the Enns Valley towards Amstetten at Hieflau with a train of iron ore from Eisenerz on 5 September 1969. No 52.798 was withdrawn in April 1973.

Below: Giesl ejector-fitted ÖBB Class 52 No 52.3512 (Krauss-Maffei No 16702, 1943) waits to leave Vienna East station with an evening commuter train on 19 July 1972. No 52.3512 was withdrawn in June 1973.

Hungary (MAV Class 520): Hungary lost all of its Class 52s in the USSR, but in 1957/58 around ten 5-foot-gauge engines were returned for cross-border work. One hundred more engines were later acquired from the USSR, all but six of which were converted to standard gauge in 1963. Four of these went to GySEV followed by four more in 1976. The class lasted into the 1980s on both railways.

Jugoslavia (JDZ (JD from 1956) Class 33): During the war 40 Class 52s were built new for SDZ in Serbia (15, as Class 33) and HDZ in Croatia (25, as Class 30, later JDZ Class 33.016 to 040). After the Jugoslavian JDZ was reinstated more engines arrived, mostly from the USSR, bringing its fleet of Class 33s to over 300 engines. In the 1960s five more came from CSD and the USSR for industrial use. Steam ended on what was by then JD in the mid-1980s, but engines in industrial use in Bosnia are still active at the time of writing (see page 36).

Luxembourg (CFL Class 56): From 1946 CFL took delivery of 20 Class 52s, ten from Belgium and ten from French manufacturers (see above). The Class was withdrawn in 1961 but four were retained to work engineering trains and were withdrawn in 1964.

Norway (NSB Class 63a): Seventy-four Class 52 engines remained in Norway after Germany's defeat. These became NSB Class 63a, although the engines retained their DRG running numbers. The last was withdrawn in 1970, although some were retained as a strategic reserve (see page 44).

Poland (PKP Classes Ty2 and Ty42): The PKP fleet was made up of some 1,200 ex-DRG engines of Class Ty2 and 150 Polish-built Class Ty42s. In 1962/63 PKP acquired a further 200 engines from the USSR for industrial use. The class lasted well into the 1990s.

Romania (CFR Class 150.1): One hundred Class 52s were delivered new to Romania in 1943/44 (CFR Nos 150.1001 to 1100). There are few later records of these engines. In 1948 the CFR received 18 additional Class 50s and 52s which it classed together as Class 150.11. Many of these came from Austria, some via Hungary and Czechoslovakia, under the auspices of the MPS. Ten further engines were received which retained their DRG numbers, but these were soon withdrawn. Both classes lasted until the mid-80s, while some remained in strategic reserves into the 1990s.

Turkey (TCDD Class 56501): During the war 43 Class 52s were leased to TCDD, while ten more, all built by Henschel in 1943, were purchased new. All 53 later became TCDD engines, Nos 56501 and 56553. They lasted until the mid-1980s.

USSR (SZD Class TE): After the war more than 2,130 Class 52s had become the 'property' of the USSR. Most of these were converted to broad gauge, and most retained their DRG running numbers, although those single engines created from several wrecked ones were numbered from 8001. In 1985, when

Graz-Köflacher Bahn (GKB) Class 52, Giesl ejector-fitted No 77 (ex-DRG No 52.6729, ÖBB No 152.6729 (WFL No 16180, 1943)), stands in the pouring rain in the yard at Lieboch Junction with a local freight on 11 July 1972. The ÖBB had reclassified its Class 52s with bar-frames as Class 152 in 1953. All were later fitted with Giesl ejectors. No 152.6729 went to GKB in 1970 and was withdrawn in 1975.

Austrian ex-ÖBB Class 52s preserved in Europe

Above left: Ex-ÖBB Class 52 No 52.7593 (WLF No 16941, 1944) is seen plinthed adjacent to the main road at Strasshof in the eastern suburbs of Vienna, Austria, on 2 September 1994. The engine is advertising 'Das Heizhaus Eisenbahnmuseum' (the 'Hot House' Railway Museum), which is situated in the old ÖBB engine shed there. No 52.7593 was withdrawn by the ÖBB in May 1976. The museum contains a large collection of steam engines associated with Austria and also houses Austria's national collection.

Above: One of the 25 engines built for the former Croatian Railways (HDZ), No 30.017 (Henschel No 27938, 1944), was once preserved at the Strasshof Railway Museum near Vienna. It was renumbered as 33.032 when JDZ was re-established after the war, and is seen here at Strasshof on 2 September 1992, but is now at the Slovak Railway Museum, Bratislava. Another of the Class 33s brought to Austria for preservation was No 33.502 (ex-DRG 52.7620 (WLF No 17591, 1944)), which was one of two Czechoslovakian Class 555s sent to Jugoslavia in the 1960s. It has since been restored in the Czech Republic and now carries its old CSD number 555.0153 again (see page 33).

computerisation of the loco stock took place, some 623 were included in a new Class 1042. Many were subsequently stored in strategic reserves across the USSR. When the USSR was dissolved in December 1991 their ownership passed to the newly created country where they were based.

Modifications and rebuilding after 1950
Many of the countries that used the Class 52s after the war carried out improvements to the engines, especially those, like Austria and Poland, that saw them as long-term investments. These included the conversion of some to oil-firing and the fitting of both Giesl ejectors and smoke deflectors. The Austerity 'Wannen' (bath tub) tenders were also strengthened, while in Austria cabins for conductors were fitted into them similar to those on the DB's Class 50s. As noted above, those built with bar-frames became ÖBB Class 152. Many of the engines taken to the USSR were converted to oil-burners, a feature they retained when eventually returned to the Soviet Bloc's allies. In the case of the 100 engines sent to Czechoslovakia in 1962/63,

oil-burning was retained and further engines converted. The latter had 3000 added to their running numbers.

The most far-reaching improvements were carried out by DR, which between 1953 and 1958 converted 25 engines to burn pulverised brown coal (or lignite), a fuel that was plentiful in the GDR. Classified 52.90, the last was not withdrawn until 1980. In addition, in 1967, 200 of the class were fitted with new welded boilers identical to those used on the Class 50.35s described above. These became Class 52.80 and lasted to the end of steam on DR in October 1988. Although officially withdrawn, 93 Class 50.80s were still listed as remaining in DB stock (recently inherited from DR) in Platform 5's 'German Railways', published in 1993. Most of these were either stored or acting as heating units.

For engines that were built as a wartime emergency measure, many of the Class 52s have had long careers. Although in Western Europe they had disappeared in the 1960s, in Eastern Europe they lasted for another 20 years. In both Poland and Turkey some were still active into the 1990s. The most remarkable of all were those sent to work the colliery lines around Tuzla, now in Bosnia Herzegovina (formerly part of Jugoslavia). During the civil war of 1992-95, which followed the break-up of Jugoslavia, several Class 52s were taken out

Above left: An IGE special leaves Rosenbach and heads for Ljubljana in Slovenia on 4 August 2002. It is headed by ex-ÖBB Class 52 No 52.7612 (WLF No 16960, 1944), which was withdrawn in January 1978 and is now preserved at Mistelbach. The second engine is ex-PKP 2-8-2 No Pt47.138 (Cegielski No 1340, 1949), now renumbered as ÖBB No 919.138.

Above: Although the last West German DB Kriegsloks were withdrawn in 1963, examples were preserved in West Germany from elsewhere. One of these, No 52.7596 (WLF No 16944, 1944), was withdrawn by the Austrian ÖBB in November 1977 and purchased by the West German Eisenbahnfreunde Zollernbahn (Friends of the Zollern Railway, EFZ) based at Tübingen in Baden-Württemburg. There it was restored for use on special trains and is seen there on Sunday 12 April 1998 during the celebrations of 25 years of steam preservation on the Zollernbahn. The engine remains at work on the Zollernbahn in 2019.

of store and returned to service. Diesel fuel was at a premium while there was a plentiful supply of coal for the steam engines. At the time of writing in 2019 some of these engines are still at work.

Preservation in Europe and Turkey

A large number of Class 52s have been preserved. The Wikipedia website 'List of Preserved Steam Locomotives in Germany' lists more than 20 original and some 100 rebuilt Class 52.80s (more than 50% of the class) in Germany alone. Two of them, Nos. 52.6666 and 52.4924, were retained by DR as active museum locos, or 'Traditionsloks'. A selection of those preserved in Germany is given in Table 1, while those preserved elsewhere in Europe are shown in Table 2.

Although some of the 'condensing' engines survived the war, two until 1964, none were preserved.

Above left: Photographed surreptitiously from a moving train between Pilsen and Prague, a Czechoslovakian Ceskoslovenske Statni Drahy (CSD) Class 555 Kriegslok was glimpsed as the train passed the depot at Zdice on 7 January 1973. From the tender it appears to have been converted to oil firing. At the time trains from Prague to Pilsen were worked by CSD Class 475 4-8-2s. After the war many of Czechoslovakia's Class 52s were taken to the USSR as 'war booty', but in 1962/63 around 100 engines were returned, giving the CSD almost 300 of the engines.

Above: Ex-CSD No 555.0153 (ex-DRG No 52.7620 (WLF No 17591, 1944)) was one of two Class 555s sent to the Kreka coal mine in Jugoslavia in the 1960s (sister engines are still at work there – see below). It is now back in the Czech Republic, via the Strasshof Railway Museum near Vienna, and restored with its original CSD number. The engine is seen at the Lužná Railway Museum on 27 June 2009. No 33.502 was bought by KHKD (Railway Historical Club, Prague) in 2001, and overhauled at ZOS Ceske Velenice in October 2001. In February 2003 it became another of the KHKD's working locomotives.

The Hungarian Magyar Allamvasutak (MAV) Class 520s

Above: In the 1960s Hungary was subject to one of the most repressive communist regimes. While queuing to cross the border into Austria on 9 September 1967 I spotted this ex-USSR/MAV Class 520 Kriegslok through a security fence and risked a photo. It is just possible to make out the small round access door fitted to these engines while in the USSR. Hungary lost all of its Class 52s to the Soviet MPS as 'war booty', but around 100 were returned in 1963 where they became MAV Class 520. Eight of MAV's Kriegsloks were sold to the Györ-Sopron-Ebenfurti Vasut Railway (GySEV). GySEV was an anachronism as it crossed the Iron Curtain, which then formed the border between Austria and Hungary, but despite this the line remained open throughout the communist period.

Above right: MAV No 520.094 (ex-DRG No 52.7781 (O&K No 14366, 1944), ex-SZD No TE 7781) came to Hungary from the USSR in 1963 but was later sold to GySEV. Here the fireman ensures that the round smokebox door is secure before the engine heads a freight from Sopron across the border into Austria on 24 May 1978. A second engine, No 520.018 (ex-DRG No 52.7443 (Skoda No 1519, 1944) ex-SZD No 7443) was at the rear of the train to provide banking assistance. No 520.094 was withdrawn in October 1979. *Brian May*

Below: One of the ex-MAV/GySEV engines, No 520.030 (ex-DRG No 52.3535 (Krauss-Maffei No 16661, 1943, ex-SZD No TE-3535) is preserved at the Narrow Gauge Railway Museum at Fertoboz, a station on GySEV in Hungary, and was pictured there on 9 October 1992. It was another of the DRG's Kriegsloks taken to the USSR at the end of the war, but returned to Hungary in 1963. In 1976 it was sold to GySEV, and was withdrawn in 1984.

The Jugoslavian Jugoslovenska Zeleznice (JDZ) Class 33s

Above: After the war Jugoslavia's JDZ eventually had over 300 working Kriegsloks on its network. Fifteen of them were built new for the Serbian SDZ (Class 33) and 25 for the Croatian HDZ (Class 30). Others had been left in the country by the retreating Germans, while more than 120 had come from the Soviet Bloc countries, many from the USSR. All of the engines became Class 33 when the JDZ was re-established after the war (it became JD in 1956). One of these, No 33.091 (ex-DRG No 52.3235 (Jung No 11246, 1944)), is seen working a stone train near Pristina (now in Kosova) on 23 August 1967.

Above right: Ex-JDZ Class 33s are still at work in Bosnia Herzegovina, in the former Jugoslavia, and are the last working Kriegsloks in the world. The collieries around Tuzla, now in Bosnia Herzegovina, were always important to the Jugoslavian economy. As a result several Class 33 Kriegsloks were drafted in to work the colliery lines. During the Civil War of 1992-95 that followed the break-up of Jugoslavia, in the absence of diesel fuel several were taken out of store and returned to traffic. One of them, No 33.248 (ex-DRG No 52.4779 (MBA No 13830, 1943)) is serviced at Bakije Mine depot on 14 June 2016.

Right: Photographed on the same day, Bakije Mine depot appears to have plenty of spare Kriegslok brake pumps, lined up for future use.

Right: No 33.504 (ex-DRG. 52.793 (Henschel No 28118, 1944)) leaves Bakije Mine depot to work an RTC special from Tuzla to Banovici on 14 June 2016. No 33.504 was one of five engines that came to the Kreka mining company at Tuzla from the USSR and Czechoslovakia in 1964. Another of them, No 33.502 (ex-DRG No 52.7620, ex-CSD No 555.0153 (WLF No 17591, 1944)) is now preserved in the Czech Republic (see above).

Below: No 33.064 2-10-0 (ex-DRG No 52.1134 (DWM/Cegielski No 548, 1943) worked around Belgrade during the war and went to the Jugoslavian JZ in 1945. Now at Tuzla, it shunts coal wagons under the coal hopper at the Sikuljie Mine, Bosnia, on 15 June 2016.

Class 33 preserved by Serbian Railways, Zeleznice Srbije (ZS)

Below right: Retained by Serbian Railways (ZS) for special workings, ex-JDZ No 33.087 (ex-52.2802 (Henschel No 28366, 1944)) heads away from Kragujevac with an LCGB/WRS special to Kraljevo on 20 September 2001.

Class 33s preserved by Slovenian Railways, Slovenske Zeleznice (SZ)

Right: For many years Slovenian Railways (SZ) has run regular public steam excursions on summer weekends. Here ex-JDZ No 33.037 (ex-HDZ No 30.022 (Henschel No 27943, 1944)) is admired by its mainly Slovenian passengers while pausing at Bled Jezero between Jesenice and Nova Gorica on 11 June 2016.

Below: Slovakia is very aware of its old status within the Austrian Empire. The 'Archduke Franz Ferdinand' of Austria, his wife 'Sophie, Duchess of Hohenberg' and a bodyguard pose in front of Kriegslok No 33.037 at Bled Jezero on 11 June 2016. The Archduke was assassinated in Sarajevo, then also part of the Austrian Empire, in 1914, an event that led to the outbreak of the First World War, although No 33.037 was built 30 years later!

The Polish Polskie Koleje Panstwowe (PKP) Classes Ty2 and Ty42

Below right: Poland's PKP was the last major user of Class 52s (PKP Class Ty2) on the main line in Europe. At their peak in the 1960s some 1,200 foreign-built Class Ty2s and 150 home-built Class Ty42s were in use, together with at least 20 more on industrial lines. Many survived into the 1990s, although by then most were stored out of use. One of the active engines, No Ty2.392 (ex-DRG No 52.3641, Krenau No 1263, 1943) is seen at Nysa Depot, situated to the south of Wrocław, in September 1984. The engine has retained its rigid-framed 'Steifrahmen' tender. It was withdrawn two years later.
Adrian Cornish

Right: Shafts of sunlight pierce the gloom inside the shed at Rozwadow in eastern Poland on 17 March 1990. The two visible Kriegsloks are PKP Nos Ty2.1010 (ex-DRG No 52.3750 (WLF No 17277, 1944)) and the 'Wannen' (bath tub) tender of No Ty2.964 (ex-DRG No 52.2848 (Henschel No 28205, 1944)).

Below: In 1990 PKP's Class Ty2s were still in charge of traffic over the mountainous line from Chabowka to Nowy Sacz. Looking to be in excellent external condition, one of the engines in use on the line, No Ty2-632 (ex-DRG No 52.6027 (BMAG No 12468, 1943)) is turned at Nowy Sacz on 13 March 1990. Five further Kriegsloks were on shed. The engine was withdrawn in July 1991.

Below right: Another PKP Class Ty2 heads a train through the snowy mountains at Męcina near Limanowa on the Chabowka to Nowy Sacz line in March 1987. *Adrian Cornish*

Right: Another outpost of steam in Poland in 1990/91 was the shed at Jaworzyna, later to become a PKP 'Museum Depot'. It retained a couple of regular turns for its Class Ty2s into the summer of 1991. Here the 'Wannen' (bath tub) tender of PKP Class Ty2.305 (ex-DRG No 52.2520 (Henschel No 27688, 1943)) is clearly seen as it leaves the small station at Borow while working the 15.16 Marciszow to Jawor train on 24 July 1991.

Below: At Jawor No Ty2-305 ran round its train, driven by me. Afterwards the PKP driver (to the right) and I clean up with hot water from the engine while the fireman looks on from the cab. *Marcin Matacz*

PKP Class Ty2s and Ty42s dumped in Poland

Right: By the 1990s derelict PKP Class Ty2s and Ty42s could be found dumped across the country, and two examples were photographed outside the PKP depot at Wrocław on 1 August 1993. On the right is PKP No Ty42.106 (Cegielski, No 892, 1946), while beyond is No Ty2.80 (ex-DRG No 52.2108, (Henschel No 26864, 1943)). Both have since been scrapped.

Right: No Ty2-1084 (ex-DRG No 52.5114 (MBA No 14318, 1944)), is viewed from the roof of Jaworzyna shed on 24 July 1991, prior to being cut up there. It had been withdrawn the previous April.

Below: Polish-built No Ty42.87 (Chrzanow No 1564, 1945) had been stored at Wolsztyn for many years, but when seen there on 1 May 2000 it was in the process of being cut up.

PKP Class Ty2s preserved in Poland

Right: Polish working Museum Railway Depots were established just before the end of Communism in Eastern Europe. The best know is at Wolsztyn, where there is a collection of historic PKP steam engines, some of them in working order. Here on 25 July 1991 PKP No Ty2.249 (ex-DRG No 52.2261 (Henschel No 27429, 1943)) was the station pilot and is seen drawing empty stock out of the station. The engine was withdrawn in July 1992. Steam remains at Wolsztyn today thanks to the British-led 'Wolsztyn Footplate Experiences', which allow enthusiasts to drive scheduled steam trains on the PKP main line to Poznan.

Left: A long-term resident at Wolsztyn has been PKP No Ty2.406 (ex-DRG No 52.4770 (MBA/O&K No 13821, 1943)). On 6 August 1993 the engine worked a pick-up goods to Slawa via Kolsko, where the train had to reverse; it is seen running round its train through the grass-covered track at Kolsko station. For a fee of $5.00 a passenger coach for enthusiasts would be attached to the train, as was the case here.

Below: The first of the Polish-built Class 52s, PKP No Ty42.1 (Chrzanow No 1506, 1945), is seen after preservation at the Jaworzyna Museum Depot on 1 August 1993.

Above: Awaiting preservation at Jaworzyna on 1 August 1993 was PKP No Ty2.223 (ex-DRG No 52.2127 (Henschel No 26883, 1943)). The engine was one of some 70 PKP engines converted to oil-firing in the mid-1960s; the tenders were adapted accordingly.

Above right: Another PKP Ty2 that was stored at Jaworzyna for a number of years was No Ty2.1035, (ex-DRG No 52.3914 (MBA No 14168, 1944)). In 2010 this engine was used as the artwork 'Train to Heaven' by the artist Andrzej Jarodzki at Strzegomski Square in Wrocław. Seen here on 15 November 2010, minus its tender, it is positioned on rails at an angle of some 80 degrees from the horizontal! *Wikipedia commons, Masur*

Right: Long before it became 'Train to Heaven', No Ty2.1035 is seen at the PKP depot at Jaworzyna on 1 August 1993.

Right: Two of the PKP's Class Ty2s were active at the Chabowka Museum Depot in the 1990s. One of them, No Ty2.953 (ex-DRG No 52.2817 (Henschel No 28163, 1944)) is seen in action on a photo charter near Rabka, on the Chabowka to Nowy Sacz line, in May 1997. The second active engine was No Ty2.911 (ex-DRG No 52.1346 (DWM No 812, 1944)). *Adrian Cornish*

Class Ty2s and 52s preserved in Belgium and Great Britain

Below left: The ex-DRG Kriegsloks in Belgium were classified Class 26, but all were withdrawn in the 1960s. An ex-PKP example was purchased by the Railway Museum at Maldegem and renumbered 26.101. After the war 100 of the engines ran in Belgium as SNCB Class 26. No 26.101 (ex-DRG No 52.3554 (Krauss-Maffei No 16691, 1943)) is pictured here at Maldegem on 4 May 1997. After the war it was taken to the USSR as 'war booty', becoming SZD No TE-3554, later going to Poland as PKP No Ty2-3554.

Below right: A Kriegslok heads a local train across the plains of Central Europe? In fact, the train is crossing Castor Meadows on its return to Wansford from Peterborough on the Nene Valley Railway in March 1991. It is headed by ex-PKP No Ty2.7173 (DRG No 52.7173 (WLF No 16626, 1943)), one of the class built for industrial use. This was the only Kriegslok to work on a heritage railway in Great Britain. This view at Castor is no longer possible as tall willow trees now form a screen along the track. No 52.7173 is now in Belgium.

The Steam Engines of World War II in Europe

Class 52s in Romania and Turkey

Left: Once Romania had become Germany's ally in 1943 it received 100 new Class 52s, which became Caile Ferate Romane (CFR) Class 150.1. The fate of many is unknown, but one of them, No 150.1088 (Henschel No 28097, 1944), had long been retired when seen dumped at Oravita on 12 May 1996.

Above: Seventy-four Class 52s remained in Norway after the war, where they became Norwegian State Railways (NSB) Class 63a. After their withdrawal, four of the class were stored in the Drangedal Tunnel, together with other NSB steam engines, until two were recovered in 1972. One of them, ex-NSB No 5865 (ex-DRG No 52.5865 (Schichau No 3063, 1944)), was subsequently preserved at the Bressingham Railway Museum in Norfolk, where it was pictured in September 1991. Named *Peer Gynt*, it is now the only Kriegslok in Britain. The other NSB engine, ex-DRG No 52.2770 (Henschel No 28322, 1944), is preserved in Norway.

Left: One of the CFR's Class 150.11 2-10-0s, No.150.1105 heads a special train, chartered by RomSteam-Aldo organisation on behalf the LCGB and WRS, away from Sibiu towards Podu Olt on 9 October 1997. The engine was one of 18 received from Eastern Block countries in 1945. It was originally planned as a basic Class 50 ÜK engine, No 50.3240 (BMAG, No 12201, 1943)) but was reclassified KDL 1, No 52.196. The second engine is CFR No 230.224, one of the many 4-6-0s built at Romania's Resita and Malaxa works based on the Prussian Railway Class P8.

Below: Romania's steam engines were renowned for the amount of information carried on their cab sides, and this is evident on immaculately turned out No 150.1105, seen after arrival at Podu Olt on 9 October 1997. The topmost plate bears the initials CFR, while below it is the plate bearing the engine's number. Below that are three plates that, from left to right, give details of the last repairs, etc, the maximum authorised speed (80km/h), and the engine's depot of Sibiu. At the time it was one of a fleet of historic CFR steam engines based there, but today they are no longer in working order.

The Turkish Turkiye Cumhuriyeti Devlet Demiryollari (TCDD) Class 56501

Above: TCDD No 56512 (ex-DRG No 52.365 (BLW No 15462, 1943), is seen stored at Halkali shed on 14 December 1973. It was one of the Class 52s which were at first hired from the DRG and later handed over to the TCDD in 1943. Like many of its class in Turkey No 56512 was not carrying smoke deflectors. Behind is 0-8-0 No 44504 (Henschel No 11334, 1912). For many years electric traction replaced steam at Halkali for the final 28km into Istanbul. In 1973 steam had recently been replaced by diesels on these trains including the Orient Express. No 56512 later moved to Asian Turkey and is now plinthed at Sultanhisar railway station.

Right: TCCD No 56504 (Henschel No 27738, 1943) was one of ten Kriegsloks built specially for Turkey, a neutral country during the Second World War, by Henschel in 1943 (Nos 56501 to 56510). It is seen waiting to leave Izmir Alsancak with a train for Buca on 2 May 1982. The engine is fitted with the smaller Witte-type smoke deflectors and had been released from Izmir Halkapinar works after overhaul in January 1982. It is now preserved at the Ankara Railway Museum. *Brian May*

Right: A train from Afyon to Izmir waits to leave Usak behind TCDD 2-10-0 No 56548 (ex-DRG No 52.7429 (WLF No 16882, 1944)) and 0-10-0 No 55016 (Nohab No 1799, 1928). No 56548 is attached to a rigid-framed 'Steifrahmen' tender. The train had arrived earlier behind another Kriegslok, No 56518 (ex-DRG 52.4857 (MBA No 13921, 1943)). *Brian May*

Below left: The shed fitters at Afyon shed seem surprised that their charge, TCDD No 56544 (ex-DRG No 52.7425 (WFL No 16878, 1943)) is attracting such attention from a group of western enthusiasts. When pictured on 5 May 1982 the engine was also attached to a rigid-framed 'Steifrahmen' tender. *Brian May*

The former Soviet Railways (SZD) 5-foot-gauge Class TEs

Below right: After the war more than 2,100 of the DRG's Class 52 Kriegsloks were taken by the Soviet MPS and classed TE (Trofeinyi 'war booty' Class E). Most of those that remained in the USSR were converted to 5-foot gauge. One of them, No TE-322 (ex-DRG No 52.322 (WLF No 9694, 1943)), now operated by Russian Railways (Rossiyskie Zheleznye Dorogi – RZhD) storms through the suburbs of Rostov-on-Don at Selmash with an LCGB special train to Novocherkassk on 21 March 2002. The engine was handed over to SZD by the MPS in 1951. Originating on the Austrian BBÖ, it also worked briefly for both MAV and CFR. *Brian May*

Another Class 52 Kriegslok to remain in the USSR was ex-SZD No TE-5653 (ex-DRG No 52.5653, (Schichau No 3931,1943). It is seen here heading a special train organised by the Dzherelo organisation for the WRS through the Carpathian Mountains in the Ukraine between Kolomyia and Rahov, on the border with Romania, on 6 October 1994. The train is crossing a viaduct near Delyatin, assisted in the rear by ex-SZD dual-unit diesel-electric No TE3.5475, a class first introduced in 1953. More than 6,800 of the latter engines were built in the USSR. No TE-5653 was then based at Osipovichi in Belarus.

2.2 The DRG Class 42 'Austerity' 2-10-0s (KDL 3)

The scale of production of Class 42s (KDL 3) was considerably less than the Class 52s (KDL 1). Despite the increasing number of Class 52s, Germany urgently needed more powerful engines to supply its troops on the Eastern Front. The design chosen was essentially a slimmed-down version of the heavy DR Class 44 2-10-0s, which were already being built as 'Übergangskriegslokomotiven' (ÜK) (see Chapter 1.2). These could be used on routes where a higher axle-load was permissible. The new engines were given a Class 52 chassis but with a shortened Class 44 boiler and two larger cylinders. The engines had a top speed of 80km/h and an axle-load of 17.6 tonnes. Stripped of all but essential parts they weighed 98.7 tonnes, some 11.5 tonnes lighter than the three-cylinder Class 44s. They were paired with either 'Wannen' (bath tub) or 'Steifrahmen' tenders as used with the Class 52s.

The first production loco, No 42.501, was built by BMAG (No 1218) and was delivered in February 1944. More than 3,000 engines were planned, although only 862 were built by BMAG, Schichau and Esslingen in Germany, and by WLF in Austria (Nos 42.501 to 42.2810) before production ceased due to the Allied invasion. By the end of 1944 many of the engines were already in the hands of the invading powers, while an order for 100 engines from DWM/Cegielski in Poland was cancelled. The last of a batch of 72 engines ordered from WLF was not delivered until 1949, Works No 2772..

KDL 3s preserved in Germany, Austria, Bulgaria and Luxembourg

Three Class 42s are now preserved in Germany. One of them is No 42.1504 (Esslingen No 4874, 1944), which latterly ran in Poland, first as PKP Class Ty3-3 then, after it had received a Polish-built boiler, Ty43.127. It is pictured at the Speyer Transport Museum on 1 October 2014. It is one of only two German-built KDL 3s to survive, the other being PKP No Ty3.2 (see page 52).

The other two Class 42 2-10-0s preserved in Germany worked on the Bulgarian BDZ as Class 16. One of these, BDZ No 16.15 (WFL No 17640, 1949) is among Bernd Falz's large private collection of engines at Hermeskeil. The other, BDZ No 16.16 (WFL No 17654, 1949) is at the Bavarian Railway Museum, Nördlingen. No 16.15 is pictured here at Hermeskeil on 13 September 2014. The locos' allocated DRG numbers, 42.2754 and 42.2768, were never used.

Post-War Distribution

After the war some 70 engines ended up in the USSR, via the MPS as 'war trophies'. These came either directly from the DRG or via the Austrian BBÖ and Hungarian MAV. In the USSR they became SZD Class TL and were withdrawn by 1962. In Germany the DB ended up with almost 550 engines, compared with around 40 on the DR. The DB engines were withdrawn by 1960, but those with the DR lasted a decade longer. Some 60 engines remained in Austria, 47 of which were built by WLF in Vienna. Most were withdrawn in the 1950s with six going to Hungary in 1957/58. Here they became MAV Class 501, some lasting until 1970.

In the post-war period the Cegielski works in Poland built 124 engines between 1946 and 1949, numbered PKP Ty43.1 to 43.124. Ty43 indicates that they were 2-10-0 freight engines built in Poland to a design of 1943. In addition to those built by Cegielski, the PKP had three German-built engines, Ty3-1 to Ty3-3, later renumbered Ty43-129/126 and 127 when they received Polish-built boilers. In addition between 1947 and 1949 the WLF Factory built 20 engines for Luxembourg, CFL Nos. 5501 to 5520, with 33 going to Bulgaria as BDZ Class 16. A further CFL engine, No 5521, was built by Esslingen (No 4873) in 1943. The class remained in

Above left: No 42.2708 (WLF No 17591, 1945) is one of the many engines built in Vienna during and after WWII. Withdrawn in 1963 this was displayed at the Vienna Technical Museum but is now at the Strasshof Museum Depot near Vienna. It is pictured on 2 September 1994 carrying its original ÖstB (Österreichische Staatseisenbahn) livery. After the war Austrian railways became the BBÖ once more (ÖBB from 1947).

Above: Two more ex-BDZ Class 16s built by Floridsdorf after the war are preserved in Austria at the ÖGEG Railway Museum at Ampflwang. One of them, ex-BDZ No 16.19 (WLF No 17636, 1949) is pictured there on 14 September 2014. The other ÖGEG engine is ex-BDZ No 16.18 (WLF No 17639, 1949). Their allocated DRG numbers, 42.2750 and 42.2753, were not used until preservation.

use on the CFL until 1964 and on the Austrian ÖBB until 1970. In Poland and Bulgaria they lasted into the late 1980s.

Preservation
Only two German-built engines survive, one in Germany, the other in Poland. Both are ex-PKP engines.

All but one of the other preserved engines were built in Austria by WLF. These include one from the Austrian ÖBB, one from the Luxembourg CFL and six from the Bulgarian BDZ. Of these, one CFL and two BDZ engines have been retained for special workings in Luxembourg and Bulgaria respectively. It appears that today only one of the BDZ engines is still active, the other being used as a source of spares.

Two of ex-BDZ locos are preserved in Germany, with two in Austria.

In addition, a Polish-built example (PKP Class Ty43) is also preserved (see Tables 1 and 2).

2.3 Other KDL and Heeresfeldbahnlokomotiven (HF) – German Army Locomotives

In addition to the large KDL 1 and KDL 3 2-10-0s described in Chapters 2.1 and 2.2 above, 11 other KDL types were identified. Apart from KDL 2, which was another large 2-10-0, the others were all small tank engines, some of them for use on narrow-gauge lines. None were built in any numbers and a few classes may not have been built at all. It is strange that the German Army saw no use for a sturdy standard-gauge 0-6-0 tank for yard shunting. Both the WD and USATC ensured that they had a large number of 'Austerity' saddle tanks and Class S100 0-6-0 side tanks at their disposal respectively (see Part 2).

Thirty-three WLF-built Class 42s went to Bulgaria after the war, where they became BDZ Class 16, which remained active into the 1980s. Two BDZ engines have been restored to work trains for tourists and enthusiasts. One of them, the first of the class numerically, No 16.01 (WLF No 17647, 1949), stands at a small wayside station with an LCGB special train from Gorna Orjahoritsa to Ruse on 4 May 2011. The coaches are from the former Royal Train of King Boris III (1894-1943). The engine's allocated DRG number was 42.2761.
Brian May

Above: The second Class 16 preserved in Bulgaria is No 16.27 (WLF No 17632, 1948). It is standing at Veliko Tarnovo prior to working a special train for the PTG, the 09.10 Veliko Tarnovo to Shumen, on 14 April 2018. This engine is now thought to be the only operational example in Bulgaria and is receiving spare parts from No 16.01. Its allocated DRG number was 42.2746. *Brian May*

KDL 3s preserved in Luxenbourg

Above right: In addition to the Floridsdorf-built Class 42s that went to Bulgaria, 20 were sold to Luxembourg's CFL and worked there until 1964. One of them, No 5519 (WLF No 17615, 1947) has been preserved, and is seen here working a special train near Schneidhain on the line between Frankfurt-Höchst and Königstein-im-Taunus in Germany on 15 May 2016. No 5519 is assisted by a DB B-B Class 211 diesel-hydraulic loco. No 5519's allocated DRG number was 42.2718. *Gregor Atzbach @ www.bundesbahn.net*

KDL 3s preserved in Poland

Right: The Class 42s also worked in Poland, and after the war they were classified PKP Classes Ty3 and Ty43. The PKP's three Class Ty3s were all built in Germany, while 124 Ty43s were built in Poland by Cegielski after the war. Wolsztyn Museum Depot has an example of each type, although neither is currently in working order. The two engines are seen at Wolsztyn on 6 August 1993. In front is No Ty43.123 (Cegielski No 1354, 1949), the penultimate one to be built in Poland after the war, while behind is Ty3.2 (ex-DRG No 42.1427, Schichau No 4448, 1944), one of three German-built machines that remained in Poland after the war.

The other KDL classes are described below based on available information, together with two classes built direct for the German Army.

KDL 2: the Czechoslovakian (CSD) Class 534.0 2-10-0s

The KDL 2 'Kriegsdampflokomotiven' were to be based on the Class 534.0 2-10-0s of Czechoslovakian Railways (CSD). These were a development of an earlier Austrian 2-8-0 and had first appeared in 1923. Construction continued with improvements to the design until the German occupation in 1939. During this period the railways of Czechoslovakia were split into three: Slovakia (SZ), the Bohemian-Moravian 'Protectorate' (CMD), and the Sudetenland, which had been absorbed into the Reich (DR). Czech territory was also lost to both Poland and Hungary.

In 1940 the two ex-Czechoslovakian loco factories, Skoda at Pilzen and CKD in Prague, were already building Class 50 and 52 engines on German command. Nevertheless CMD and SZ somehow persuaded the authorities that these were not universally suitable for their countries' railways. The construction of ten engines was therefore authorised by the Germans, five at each factory. Although they were built to a modified design no simplification was involved and it is therefore unlikely that they were what the Germans had in mind for their Class KDL 2. The engines weighed 84.0 tonnes and could be identified by their three domes, compared with just two on both the earlier and later engines. Smoke deflectors were also added for the first time, in contrast to the KDL 1 and KDL 3 engines, where they had been dispensed with. The ten engines were initially numbered from 534.0124 on each railway, but after the war they became CSD Nos 534.0139 to 0149 and worked until the early 1970s. The design was then revised yet again, and by 1947 more than 200 new engines, Class 534.03, had been built by Skoda and CKD.

Preservation: None of the war time engines have survived, but examples of both the earlier, No 534.027, and later versions have been preserved. Details are given in Table 3.

Right: The museum status of Wolzstyn shed often meant that larger engines than were really required were used. On 25 July 1991 No Ty3.2 was turned out to work the midday train from Wolsztyn to Nowa Sol and Nowe Miasteczko, which comprised just two coaches. Here No Ty3.2 is seen approaching a bridge over the main road near Lubiecin en route to Nowa Sol on 25 July 1991; overbridges are unusual around Wolsztyn, as the land is flat, and only level crossings are needed.

Right: No Ty3.2 is seen after arrival at Nowe Miasteczko. The engine has run round its train and its fireman can be seen jumping off the front buffer-beam ready to couple the engine to its two coaches.

Right: **Type KDL 2** was to be based on the Czechoslovakian (CSD) Class 534.0 2-10-0s, but it appears that the idea was not pursued. Nevertheless during the war the then separate Railways in Slovakia (SZ) and Bohemian-Moravian 'Protectorate' (CMD) persuaded the Germans that more of the class were needed, and as a result five more were built by Skoda and CKD for each railway. After the war the design was revised again, and more than 200 Class 534.03s were built for the re-established CSD. One of the later engines, No 534.0323, is seen at a suburban station on the outskirts of Prague on 6 January 1973. It is fitted with a Giesl ejector and coupled to one of the later, higher-sided tenders, No 935.117.

Left: Very few ex-CSD Class 534s have been preserved. They include, however, one of the pre-war engines and three of the later Class 534.03s, but none of the KDL engines. One of the later engines, CSD 2-10-0 No 534.0432 (CKD No 2335, 1946), is seen at the Luzna Railway Museum in the Czech Republic on 27 June 2009. The engine is also fitted with a Giesl ejector. At the time the engine was kept at the Czech Railways (CS) depot at Klatovy and used for special workings.

KDL 4: 0-8-0 industrial tanks

The 'ELNA Type 6' was the last of six classes of standard-gauge German tank engine, identified by the ELNA (German Locomotive Standards) Committee after the First World War, to modernise Germany's industrial railways. Examples were built by BMAG and Henschel both before and after the war as either side or well tanks. The design was selected for construction as KDL 4 and an order for 70 engines was placed with Henschel in Germany, but this was later transferred to the French firm of Schneider. Only three are thought to have been completed before France's liberation. However, 67 well tanks were later constructed by Schneider and used by SNCF as its class 040TX. The fate of the other three is uncertain, but they are thought to have been destroyed. SNCF's last KDL engine was withdrawn in 1971.

Preservation: One pre-war and one post-war example are preserved in Germany (see Table 3). None of the SNCF examples survive.

KDL 5: 0-10-0 Henschel industrial tanks

KDL 5 is described as being a Henschel design, possibly for use at the Homécourt steelworks in north-east France. No other information has been found.

KDL 6: 0-8-0 industrials

KDL 6 is described as being a Henschel design to be built by Batignolles in France. No other information has been found.

KDL 7: 0-8-0 industrial tanks

KDL 7 is described as being an O&K (MBA) design to be built by Franco-Belge in Belgium. No other information has been found.

One of the KDL 4 0-8-0 well-tanks, No 040TX-43, built by the French firm of Schneider in 1944, is seen shunting near Boissy-Saint-Léger in the Paris suburbs on 29 March 1969. The last example was withdrawn in 1971 and, although none of these engines were preserved, two of the German-built examples were more fortunate. (see Table 3.)

KDL 8: 0-4-0 industrial tanks

The KDL 8 0-4-0 tanks were the smallest of the standard gauge industrial KDL tank engines. The WLF works in Vienna was given an order to build 11 but by 1944 the Axis powers were losing the war and only seven were constructed. These worked at the Schoeller-Bleckmann Steelworks situated to the south-west of Vienna.

Preservation: At least three are preserved, two in Germany and one in Austria (see Table 3).

KDL 9: 900mm-gauge 0-6-0 industrial tanks

KDL 9 is described as being a Henschel design possibly for use at the Harcourt steelworks in north-east France. No other information has been found.

KDL 10: 900mm-gauge 0-4-0 construction tanks

The KDL 10s were built pre-war to the Henschel 'Klettwitz' design. Nine were built by Krauss-Maffei. A superheated version was built after the war.

Preservation: One has been preserved (see Table 3.)

Right: The first of the seven **KDL 8** 0-4-0 tanks which worked at the Schoeller-Bleckmann Steelworks (SBS No 01 (WLF No 16111, 1944)), is now preserved at the OeGEG Museum Depot at Amplfwang in Austria. It was pictured there on 14 September 2014.

Below and below right: Two further ex-Schoeller-Bleckmann steel works KDL 8s are preserved in Germany. One of them, SBS No 02 (WLF No 17324, 1944), is seen in the first picture at the private railway museum at Hermeskeil on 13 September 2014. A third, SBS No 04 (WLF No 17323, 1944), is preserved in Germany at the Transport Museum at Prora on the island of Rügen, the subject of the second view, dated 16 September 2010.

KDL 11: 600/750/760mm-gauge 0-8-0 industrial tanks, German Army (Heeresfeldbahnlokomotiven (HF. Class 160D))

These engines were built in Belgium by Franco-Belge for the German and Austrian 750mm and 760mm systems. They often worked attached to a small tender. After the war a number were used in Austria on ÖBB's extensive 760mm network, and on the Salzkammergut-Lokalbahn (SKGLB) from Salzburg. The ÖBB engines became Class 699.0, while those that were rebuilt with extended tanks and ran without a tender became ÖBB Class 699.1.

Preservation: Nine have been preserved, including one in Wales (see Table 3).

KDL 12: 600/750mm-gauge 0-6-0 tank engines, German Army (HF Class 70 C)

In 1939 Henschel built two prototype engines for the 600mm gauge. A larger version for the 750mm gauge with an improved boiler was designed in 1940, but not built. In 1941 Henschel built 17 engines to the original design but using the improved boiler.

Little is known about these engines although one is reported to have worked at Znin in Poland as PKP Ty1-62 (Henschel No 25961, 1941), but this was scrapped in 1957. Znin is now part of the PKP's museum containing its 600mm engines at Wenecja.

Preservation: None of the class is thought to have survived.

Above: The Class **KDL 11s** were built in Belgium by Franco-Belge for the German and Austrian 750mm- and 760mm-gauge networks. Nine examples have been preserved, including this engine (Franco-Belge No 2836, 1944, HF No 2836), which retains its tender. It is now preserved at 'Le P'tit train de la Haute-Somme' (CFCD), where it runs as the line's No 10. It is seen heading away from Port Cappy, on the Canal du Somme, heading for the CFCD terminus at Dompierre on 18 May 1998. Supplies were transferred from barges at Port Cappy during World War 1.

Above: After the war a number of the engines were used on Austria's Salzkammergut-Lokalbahn (STLB). The line's No 699.01 (Franco-Belge No 2855, 1944, HF No 2855) was one of the class built with extended side tanks, thus dispensing with a tender. It is seen here soon after its arrival at Llanfair on the Welshpool & Llanfair Railway in Wales in 1970. *Barry Payne*

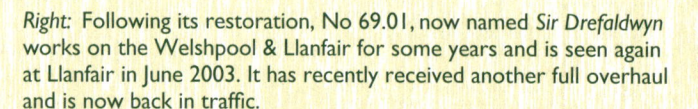

Right: Following its restoration, No 69.01, now named *Sir Drefaldwyn* works on the Welshpool & Llanfair for some years and is seen again at Llanfair in June 2003. It has recently received another full overhaul and is now back in traffic.

KDL 13: 600/750mm 'Riesa'-type construction tanks (HF Class 70 PS)

A large number of these useful 0-4-0 tanks were built by Henschel for work on construction sites before the war, followed by 34 for the HF during the war. In addition, at least 18 KDL engines were built by the Budich works in Wrocław, Poland (Breslau when controlled by the Germans) and classed PKP T4. Henschel also produced two other types of 600mm-gauge 0-4-0T for the German Army: a smaller 'Monta' type (HF Class 60PS), classed PKP T2, and a larger 'Fulda' type (HF Class 90 PS).

These engines were used throughout the territory occupied by the Germans. This included Guernsey and Jersey in the Channel Islands, the only place where German Army engines worked on British soil during the war. Most were returned to the European mainland in 1943, when construction of the Islands' vast fortifications was completed. Others remained on the Islands until they were liberated in May 1945.

Preservation: A number of 'Riesa'-type engines are preserved, a selection of which are listed in Table 3. These include four of the Budich-built KDL engines.

Other 'Army Field Railway Locomotives' – 'Heeresfeldbahnlokomotiven' (HF)

A number of narrow-gauge engines, Heeresfeldbahnlokomotiven (HF), were built directly for the German Army during the Second World War. The total was, however, small when compared with those produced during the First World War. Two examples are described below.

750/600mm-gauge 0-6-0 'tanks', HF Class 110 C

More than 130 engines of this type were built during the war by Henschel, Jung, Krenau/Chrzanow and CKD, Prague, with production continuing after the war. They often worked attached to a small tender.

Preservation: Three of the class have been preserved (see Table 3).

Three KDL 11s are preserved in Austria, including No 699.103 (Franco-Belge No 2821, 1944, HF No 2821). This engine was also rebuilt to run without a tender and was a long-term resident on the ÖBB's Steyrtalbahn. Now a heritage railway, operations are based on Grünburg where No 699.103 was pictured on 19 September 2014. As can be seen, the engine is fitted with a Giesl ejector

Large numbers of **KDL 13** narrow-gauge 0-4-0 'Riesa'-type tanks were built by Henschel both for use on construction sites and by the German Army (HF Class 70PS). The KDL versions were, however, built by Budich in Poland in 1943/44, three of which have been preserved. One of the later Henschel engines (Henschel No 28514, 1949) is seen at the German Technical Museum, Berlin, on 17 June 1999.

750/760mm-gauge 0-10-0 'tanks', HF Class 210 E

These were the largest of the HF narrow-gauge engines. A single engine with tender was built by BLW in 1939, but the firm then withdrew from building steam engines for the German Army. Five more were built by Henschel in 1944.

Post-war: The unique Borsig engine, HF No 191 (Borsig No 14806, 1939), was sold to the 760mm-gauge Austrian Salzkammergut-Lokalbahn (SKGLB) in 1945 and worked there until 1957. It was then bought by the Zillertalbahn where it was named *Castle Caereinion*, as the railway was then twinned with the Welshpool & Llanfair Railway. It has since moved back to Germany.

Preservation: Two engines have been preserved, HF No 191 and one of the Henschel examples (see Table 3).

Left and below: Three **HF Class 110C** 0-6-0 tanks built in Germany for the army have been preserved. One of them, ex-DRG No 99.4652 (Henschel No 25953, 1941), is displayed with its tender in a DR goods yard setting at Putbus on the Rügensche Kleinbahn (RüKB) on 18 September 2010. It first arrived on the Island in 1965 when the 750mm-gauge line was part of the DR.

The most powerful of the narrow gauge HF engines were the heavy **Class 210E 0-10-0Ts**. One of them, HF No 191 (BLW/Borsig No 14806, 1939), was sold to the 760mm-gauge Salzkammergut-Lokalbahn in 1945 and worked there until 1957. It was then bought by the Austrian Zillertalbahn, where it ran as No 4. It is seen pictured right at Jenbach on 2 September 1969, named *Castle Caereinion*, as the railway was then twinned with the Welshpool & Llanfair Railway. It was sold to a private buyer in 1972 before moving to the 750mm-gauge RüKB line from Göhren to Putbus on the island of Rügen, the 'Rasender Roland' ('Raging Roland'). The engine is seen in the second picture at RüKB's Putbus depot on 20 September 2010, named *Aquarius C*. In July 2018 it returned to Austria to work on the 760mm-gauge Taurachbahn at Mauterndorf.

Part 2 British and American War Engines   Introduction

Introduction to Chapters 3 and 4

Like the Germans, both the British and Americans had recognised the importance of railways during the First World War. As a result, in the late 1930s, as war clouds gathered once again, contingency plans were made to produce the necessary locomotives and rolling stock to supply an army in Continental Europe. Initially, however, as America was unwilling to get involved in another European war, this mainly involved Great Britain.

The British War (WD) Engines

The body responsible for ordering the equipment necessary to run military railways both in Britain and overseas was the Ministry of Supply (MoS), while the War Department (WD) organised their deployment. The engines were not only needed to support an army in the field but were also required in the many supply depots, docks and airfields that were built at home. Many of the troops operating the military trains were recruited from the main British railway companies. The Longmoor Military Railway (LMR) near Liss in Hampshire opened in 1933 to train these men in the military operations railway.

When war was declared in August 1939, however, the best that could be mustered to support the British Expeditionary Force (BEF) heading for France was a bunch of mainly ancient engines dating from before the Grouping of railway companies in 1923, many of which had already seen military service during World War 1. In all, 79 GWR 'Dean Goods' and eight LMS 0-6-0T 'Jintys' were sent to France with the BEF. It was not until December 1939 that Stanier's 8F 2-8-0 design for the LMS was chosen as the standard WD heavy freight engine. It was expected that they would play the same role in Europe as Robinson's Great Central Railway 2-8-0s (later LNER Class O4) had during the First World War.

The evacuation from Dunkirk in May and June 1940 and the capture of the WD engines already in France ruled out any land-based military action on the European mainland for the indefinite future. (Five of the 'Jintys' survived, becoming SNCF Class 030TW until returned to BR in 1948, where they lasted into the 1960s. Many 'Dean Goods' also survived the war, but were soon withdrawn, although some went to China). The first of the WD 8Fs, delivered from May 1940, were therefore unwanted by the WD, so were lent to the 'Big Four' railway companies, then under the control of the Government's Railway Executive Committee (REC). The engines were later deployed in the Middle East.

By 1942 three British classes of 'Austerity' engines were being planned by R.A. Riddles, who in 1939 had moved from the LMS to become Director of Transportation Equipment for the MoS. Like the German 'Kriegsloks', these were designed to be built quickly using the minimum amount of labour and materials. The first 'Austerity' was a simplified version of an existing Hunslet industrial 0-6-0 saddle tank (the design was chosen instead of further 'Jintys'). The second was an 'Austerity' version of the 8F, the WD 2-8-0, while the third was an enlarged version, the WD 2-10-0. All were built by private manufacturers.

In addition, the MoS ordered a large number of Stanier 8Fs from the works of Britain's 'Big Four' railway companies for home use.

The WD, which became part of the Ministry of Defence (MOD) in 1964, renumbered its steam engines several times (see Appendix 1). Several liveries were used, ranging from black, grey, khaki green and dark green, while desert-sand was used for those destined for the Middle East. In the 1960s the remaining WD/MOD steam engines were painted an attractive, fully lined-out Prussian Blue.

One further class of steam engine is also included here: Bulleid's 'Austerity' 0-6-0 of 1942 for Britain's Southern Railway. Although not a military engine, it epitomised the 'Austerity' concept and contributed to the war effort through the transportation of troops and materials in southern England.

The United States of America's War (USATC) Engines

In December 1941 Germany declared war on the USA, to which the USA immediately responded. Once committed, huge resources became available to the Allies. The two types of steam engine of particularly relevance to the war in Europe were the Class S100 'USA' 0-6-0 tanks and the Class S160 2-8-0s. These had been designed, by Major J.W. Marsh and Col Howard G. Hill of the US Corps of Engineers respectively, to fit as many of the world's loading gauges as possible, including Britain's. The Engineers Corps was replaced by the United States Army Transport Corps (USATC) in July 1942. Most of the S160s were in USATC grey livery, while the S100s carried USATC black. Details of their numbering are

given in Appendix 1.
The operational side of the USATC, the Military Railway Service (MRS), was organised into Grand Divisions (RGDs), which in turn included up to five Railway Operating Battalions (ROB) and one Shop Battalion (RSB). The soldiers for each RGD were recruited from a single American railway company. Europe was divided into five regions, RGDs Nos 706 to 710; for example, RGD No 707 was to commence operations at Omaha Beach after D Day and was then to progress through Belgium into Germany. Its complement was recruited from the USA's Southern Railway.
The operational side of the USATC, the Military Railway Service (MRS) was organised into Grand Divisions (RGDs), which in turn included up to five Railway Operating Battalions (ROB) and one Railway Shop Battalion (RSB). The soldiers for each RGD were recruited from a single American railway company. Europe was divided into five regions, RGDs Nos. 706 to 710. RDG No 707 for instance was to commence operations at Omaha Beach after 'D' day and was then to progress through Belgium into Germany. Its complement was recruited from the USA's Southern Railway.

Allied steam engines at the end of the war

Both the Stanier 8Fs in the Middle East and the first of the British and American engines to arrive in Europe, either across the Channel or across the Mediterranean to Italy, were to see a great deal of action. Others arrived too late for the conflict and were made available to the liberated countries, particularly France, Belgium and the Netherlands. The remainder were put into store either in Britain or across the Channel.

Three of the War Engines described in the book were pictured on the back cover of this railway magazine: a Stanier 8F, a Bullied C1 and a USATC S160. The other engines are a GWR Collett Class 28XX 2-8-0, the LNER's Thompson B1 4-6-0 'Springbok' and a SR Raworth & Bulleid Class CC electric locomotive. 'All 'Pulling their weight towards Victory. *Reproduced with the permission of Mortons Media Group*

3 British War Department Steam Engines

The post-war period and preservation

After the war, although some WD engines remained in use in the Middle East, most were repatriated and sold to the 'Big Four' railway companies (and British Railways (BR) from 1948). Others were sold for industrial use, including the newly created National Coal Board (NCB). A number remained on the Continent, notably the 2-10-0s, which worked in the Netherlands and Greece. Most of the USATC engines were sent to Eastern Europe under the United Nations Relief & Rehabilitation Administration (UNRRA); this body was set up by the newly created United Nations to assist in the recovery of countries affected by the war. Some of the 0-6-0 tanks remained in Western Europe and worked for the French SNCF and for BR's Southern Region.

3.1 The Stanier 8F 2-8-0s

Construction and distribution

In December 1939 Stanier's 8F 2-8-0 design for the LMS of 1935 was chosen as the standard WD heavy freight engine. It was expected that the engines would play a similar role in Europe to that of Robinson's Great Central Railway 2-8-0s (LNER Class O4) in the First World War. The engines weighed 72 tons 2 cwt in working order with a tractive effort of 32,440lb. The initial order was placed by the Ministry of Supply (MoS) in December 1939; this was for 240 engines, to be modified for military use, sourced from three private manufacturers: the North British Locomotive Co Ltd (NB) and Beyer Peacock & Co (BP), 100 each, and the Vulcan Foundry (VF), 40. The deployment of

British Locomotives for the Middle East

A number of the Stanier Class "8F" L.M.S.R. 2-8-0 locomotives, which were adopted some time ago as the standard design for the War Department, are being built to the order of the Ministry of Supply. For service in the Middle East 50 of them, together with 92 L.N.E.R. Class "O4" 2-8-0s (the R.O.D. standard in the last war) are being equipped for burning oil, instead of coal. Those destined for working on the Persian railways will be equipped with three separate braking systems, Westinghouse and automatic vacuum on the train, and steam on the engine. Fenders at the front end, and masked headlights, are also among the special fittings, which may be observed on the L.M.S.R. type locomotive in the lower picture. Both engines have 4 ft. 8 in. driving wheels and the tractive efforts are 32,438 lb. (L.M.S.R.) and 31,326 lb. (L.N.E.R.).

Extract from *The Railway Magazine* for January 1942 regarding the Stanier 8Fs and the ex-GCR/LNER Robinson Class O4 2-8-0s. *Reproduced by permission of Mortons Media Group*

the engines in Europe was, however, aborted after the evacuation of Allied troops from the beaches of Dunkirk at the end of May 1940. As a result NB's order was reduced to 60 and BP's to 40, while the VF's order was cancelled altogether. Many of the newly built engines were sent to work for the Railway Executive Committee (REC) on British main lines. Having been driven out of Europe, the war effort was transferred to the Middle East. Although it is outside the scope of this book, most of these engines were sent to the Middle East including Persia (now Iran), Iraq, Palestine (much of which is now Israel) and Egypt. Some of them were built for oil-firing. This development was reported in The Railway Magazine of January 1942, as well as the fact that 92 of the LNER's Class O4 2-8-0s were to be converted to oil-firing for use in the Middle East.

In addition, 22 new 8Fs were sent to Turkey in 1941 as a goodwill gesture. Turkey remained neutral throughout the war and also received steam engines from Nazi Germany (see Chapter 2.1). These engines were built with right-hand drive, unlike the British versions, which were all left-hand drive. In the event, seven of these were lost at sea, three due to a collision between two ships in the Atlantic west of Ireland and four more when their ship was torpedoed. Five replacement engines were sent, three in 1941 and two more in 1943. It was not uncommon at the time for engines destined for Britain's railways to be diverted to make up numbers when WD engines were lost at sea. From 1943, as far as WD engines are concerned, the emphasis switched to the production of the 'Austerity' 2-8-0s, which are described in Chapter 3.3.

At this time the MoS ordered more engines for use at home by the REC. These

L.M.S.R. class "8F" standard 2-8-0 built recently at Doncaster, L.N.E.R.
Locomotives of this type are now being built by all the four main-line railways with L.M.S.R. numbering. The maker's plate on this particular locomotive bears the inscription "Built by and on loan to L.N.E.R. Doncaster No. 1956, 1943"

Many of the L.M.S.R. class "8F" engines are now serving overseas; above is one of them as W.D. No. 9338 at El Alamein

Extract from *The Railway Magazine* for September and October 1943 regarding the Stanier 8Fs built for the home railways and for use in the Middle East. LMS No 8510 was built at Doncaster in 1943 for home use while WD (9)338 (NB No 24538, 1941) worked in Egypt. 9000 was added to its WD number to avoid confusion with Egyptian State Railway's engines. *Reproduced by permission of Mortons Media Group*

included engines from the works of all 'Big Four' railway companies (see the accompanying extract from The Railway Magazine). In all, between 1941 and 1946 849 Class 8Fs were built for both the WD and REC by the following manufacturers: NB (208), Crewe (136), Brighton (93), Swindon (80), Horwich (75), VF (67), Darlington (53), Doncaster (50), BP (50), Eastleigh (23), and Ashford (14). Later in the war 24 of the Brighton-built and 42 of the Doncaster-built engines were purchased by the LNER, becoming Class O6. In BR days they eventually became Nos 48705 to 48772 and moved away from the BR's Eastern Region. In the case of engines pictured working for BR in this volume, a brief account of their deployment in the Middle East is given in the relevant caption.

The engines' only encounter with the European mainland, apart from possibly in Turkey, occurred during the winter of 1944/45 when 15 from the Middle East were overhauled and sent to Italy. There the engines were converted to oil-burning and became FS Class 737 Nos 001 to 015; they worked in Italy, mainly in the south of the country, until 1956.

Post-war distribution

Ninety-five 8Fs remained in the Middle East in the autumn of 1946, but a year later traffic had decreased and surplus engines were offered to the LMS. After some haggling, 39 were returned to Britain and went to Crewe to be overhauled. Ten of these had previously carried LMS numbers between 8045 and 8061, to which the BR '4' prefix was added in 1947. Those that had gone direct to the WD were renumbered 48246 to 48263 and 48286 to 48297 (apart from No 48293, which had remained on the LMS since 1940).

In 1952 five more 8Fs (WD Nos 307/20 and 508/75/83) were returned to Britain and, after a protracted stay at Derby Works, were returned to the WD, renumbered 500/1/8/11/12 respectively,

8Fs built by Beyer Peacock for the WD working for BR

WD 8F 2-8-0 No (70)438 (BP No 7018, 1941) is seen at Crewe Works soon after arrival back from Iran on 28 August 1949. It was one of only 15 engines built for the WD from the order for 100 received by BP in December 1939. After conversion to oil-firing, the engine was one of 155 sent to Iran in September 1941, 12 of which were lost en route. In Iran it became IrSR No 41.179. In February it was transferred to Palestine, but spent most of its time in store at Azzib, then at Suez in Egypt. It was one of 39 engines selected by the LMS for use at home towards the end of 1947. It had returned to Britain by June 1948 and, after rebuilding at Crewe during 1949, became BR No 48290 in September 1949. *Frank Hornby*

for work at home. They were sent to the Longmoor Military Railway (LMR), the WD facilities at Bicester (No 500), and Cairnryan Military Port, near Stranraer (Nos 508 and 511). During 1956 No 512, then an oil-burner, was involved in a fatal accident on the LMR. Despite this it was repaired and sold to BR in 1957 together with Nos 500 and 501. After overhaul all three went to Polmadie shed in Glasgow (66A) as BR Nos 48773, 48774 and 48775. No 48774 was withdrawn in July 1965, but Nos 48773 and 48775 went south and were allocated to Rose Grove shed (10F, ex-24B) and Lostock Hall shed (10D, ex-24C) in

Preservation in Europe and Turkey

The only 8F to be purchased direct from BR for preservation was No 48773 (ex-WD No 500). Fortunately three of those ordered by the MoS for the home railways – Nos 48431 (Swindon, 1944), 48518 (Doncaster, 1944) and 48624 (Ashford, 1943) – went to Barry scrapyard and have subsequently been preserved (the boiler of No 48518 is to be used in the construction of the new GWR 'County' Class 4-6-0 No 1024 at Didcot). Details are given in Table 4. Three other engines from Barry scrapyard, which were built by the LMS, have also been preserved (see Table 4a).

In addition, two of the engines exported to Turkey have been repatriated while others are preserved elsewhere (see Table 4). It is possible that there are other examples lying derelict in Turkey and the Middle East.

Lancashire in July 1968. They both survived until the end of BR steam in August of that year. Both the Cairnryan engines were scrapped in 1959.

Towards the end its career with BR, 8F No 48290 is seen passing Cardiff General with a freight on 4 November 1964. From 1957 to 1962 it was at Northampton shed (2E), then Birkenhead (6C/8H), and was withdrawn in August 1965. *Colour Rail*

BR 8F 2-8-0 No 48287 is seen on Crewe South shed (5B) on 11 June 1967, but was withdrawn before the end of the month. The engine was built as WD No (70)402 (BP No 6982, 1940), but at the end of 1940 it was not required by the WD and was loaned to the LMS, becoming No 8288. It was recalled in September 1941 and sent to Egypt. Like No (70)438, by the autumn of 1946 it was stored at Suez and was also one of the engines selected by the LMS at the end of 1947. After rebuilding at Crewe during 1949, it was renumbered as BR No 48287. *Frank Hornby*

8Fs built by North British for the WD working for BR

Like BR No 48250, No 48773 was one of the first 60 8Fs built by North British (No 24607, 1940) when it became WD No 307 (later No (70)307). It was one of the last to return to Britain. It worked first for the LMS as No 8233, before being sent to the MEF, then it too worked in Palestine, then Egypt, where it remained until 1952, It returned to Britain in July and, after overhaul at Derby, worked for the WD, renumbered 500, until purchased by BR in 1957. It was allocated to Polmadie shed, Glasgow (66A), as No 48773. The engine is seen here just after being allocated to Rose Grove shed in Lancashire, working a coal train towards Preston on 9 July 1968, just a month before the end of steam on BR.

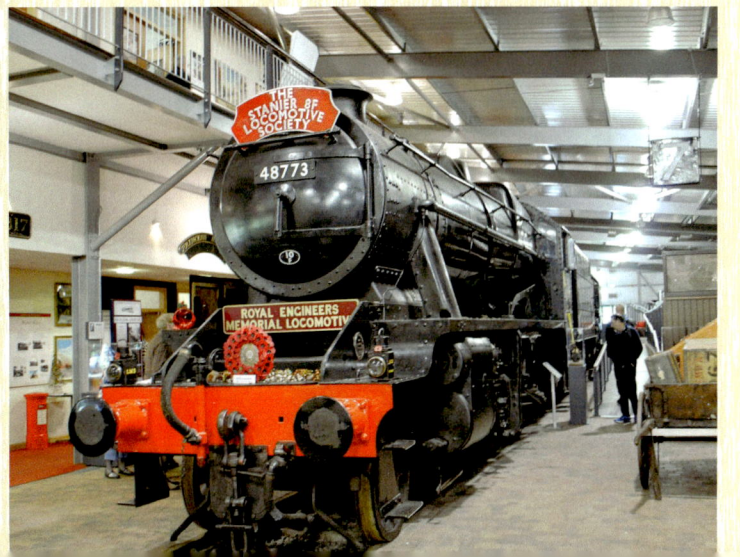

Above left: No 48773, seen in the previous picture when working for BR, was the only Stanier 8F to be purchased for preservation direct from BR, by the Stanier 8F Locomotive Society. It is seen here under restoration at Bridgnorth in June 1969 and has remained on the SVR ever since.

Above: For a time the engine ran on the SVR with its original LMS number, 8233, and is seen here drifting into Ardley station in April 1989 with a demonstration freight for Bewdley. At Ardley the freight was scheduled to pass a passenger train from Kidderminster to Bridgnorth.

Left: At present No 48773 is displayed at the SVR's Engine House museum at Highley. Seen here on 28 October 2017, it carries a Stanier 8F Locomotive Society headboard and a plaque stating that it is the Royal Engineers Memorial Locomotive.

8F ordered by the LMS in 1941, working for BR

Above: BR 8F 2-8-0 No 48206 had just been withdrawn when seen at Liverpool's Speke Junction shed (8C) on 26 May 1968. Both the MoS and LMS ordered new engines from NB in 1941; LMS No 8206 was one of the latter and was delivered in July 1942. The star on its cabside indicates that it has modified balance weights, making it suitable for fast freight duties. *Frank Hornby*

BR 8Fs ordered from various works by the MoS for the REC for home use, working for BR

Above right: No 48632 was one of the 93 engines built in Brighton during 1944, and is seen at Westerleigh with a southbound freight on 28 June 1965. It is waiting on the double track spur from the ex-LMS Gloucester to Bristol main line to reach Stoke Gifford yard, Bristol, on the former GWR Swindon to South Wales line.

Right: Two Stanier 8Fs are seen at Lostock Hall shed (10D) on 23 March 1968. That on the left is No 48476, one of 80 engines built at Swindon between 1943 and 1945, while to the right is No 48132, ordered by the LMS and built at Crewe in 1941. Lostock Hall shed was among the last to retain steam on BR and No 48476 lasted to the very end in August 1968. No 48132 was withdrawn two months earlier. *Ray Ruffell, Slip Coach Archive*

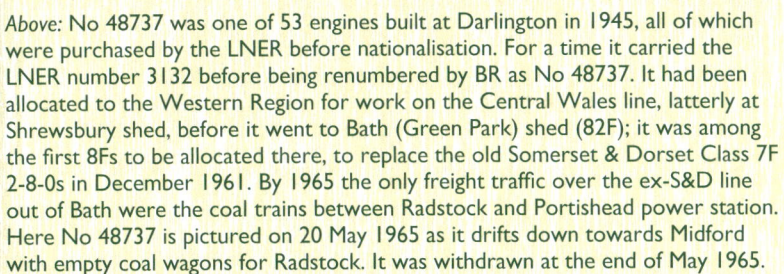

Above: No 48737 was one of 53 engines built at Darlington in 1945, all of which were purchased by the LNER before nationalisation. For a time it carried the LNER number 3132 before being renumbered by BR as No 48737. It had been allocated to the Western Region for work on the Central Wales line, latterly at Shrewsbury shed, before it went to Bath (Green Park) shed (82F); it was among the first 8Fs to be allocated there, to replace the old Somerset & Dorset Class 7F 2-8-0s in December 1961. By 1965 the only freight traffic over the ex-S&D line out of Bath were the coal trains between Radstock and Portishead power station. Here No 48737 is pictured on 20 May 1965 as it drifts down towards Midford with empty coal wagons for Radstock. It was withdrawn at the end of May 1965.

Above: Another S&D-based 8F, No 48660 – one of 23 engines built at Eastleigh Works – heads a loaded Radstock to Portishead power station coal train as it approaches Midford Viaduct on 20 May 1965. The engine came to Bath (Green Park) shed from Shrewsbury in February 1962 and was withdrawn in July 1965.

Above: No 48624 was one of 14 engines built at Ashford Works in 1943. After the end of steam six 8Fs went to Woodham's Barry scrapyard in South Wales, most of which were subsequently preserved; No 48624 was one of them, and is seen at Barry in July 1969. The engine was withdrawn in July 1965 after spending 22 years at just one shed – Willesden (1A) – arriving at Barry three months later. It left again after 16 years in July 1981, initially going to the Peak Rail project at Buxton. It has since been restored to steam.

Above right: More than 20 years were to elapse before No 48624 would steam for the first time. It is seen here, as LMS No 8624, soon after restoration as it waits at Kingsley & Froghall station heading for Cheddleton on the Churnet Valley Railway on 18 July 2010. The engine appears to be masquerading as a 'Jubilee' 4-6-0 as it is carrying the LMS passenger maroon livery.

Right: After a move to the GCR at Loughborough, the engine received its correct BR livery and was renumbered 48624. It is seen at Loughborough on 26 January 2018. The engine is currently awaiting another overhaul at the GCR.

Right: This view of TCDD No 45161 at Irmak was taken from the open corridor connection of a passenger coach on the same day. The typical 'Stanier' smokebox carried the number of another Stanier engine, 'Black 5' No 45161, which was withdrawn by BR back in November 1966.

8Fs built by North British for the WD, working for the TCDD in Turkey

Above: A work-worn Stanier 8F 2-8-0, TCDD No 45161 (ex-WD No 362/522 (NB No 24670, 1941), one of the 20 that eventually reached Turkey from Britain from 1941, is seen at work as the station pilot at Irmak, Turkey, on 17 December 1973. It was one of the five engines sent to Turkey to replace seven that had been lost at sea. The class remained active in Turkey until the end of steam there – Turkish railwaymen referred to them as 'Churchills'.

Right: TCDD No 45161 is seen again at Irmak almost ten years later as it prepares to shunt a long freight on 8 May 1982. In the background is a Class 56301 'Skyliner' 2-10-0, TCDD No 56318 (VIW No 4807, 1948), which has brought the freight into Irmak; it was one of 88 engines built in the USA for Turkey after the war. *Brian May*

TCDD No 45170 (ex-WD 554 (NB No 24755, 1942)) is seen at rest between duties at Samsun docks on 12 May 1982. Like No 45161 it was one of the seven engines sent to Turkey to replace those lost at sea, being one of the last two to be sent in 1943. It is also one of two Turkish engines, the other being No 45160, to return to Britain for preservation and both are pictured opposite. In the background is TCDD No 3302 (Henschel No 15637, 1918), one of ten 0-6-0 tanks ordered by the Turkish Controller of Military Railways during the First World War. *Brian May*

Above left: One of the Turkish engines to be repatriated to Britain is ex-TCDD 2-8-0 No 45160 (ex-WD No 348 (NB No 24648, 1941)), which has since been restored to steam. It is seen at work at Toddington on the Gloucester Warwickshire Steam Railway on 26 December 2011. The engine ran for some time as LMS No 8274 before it was exported to Turkey. *Virginia Rowan*

Above centre: Details of the cab of No 45160, complete with the Turkish crescent and star, are seen at Toddington in December 2011. These engines were delivered new from Britain with their TCDD insignia already attached. No 45160 is now at Ruddington on the GCR(N). *Virginia Rowan*

Right: Ex-TCDD No 45170 (ex-WD No 554 (NB No 24755, 1942)) is seen again, now back in Britain inside the 'Locomotion' Museum, Shildon on 28 January 2011. It arrived at Portbury Docks from Turkey during December 2010. *Steve Frost*

Top far right: A close-up of the Westinghouse brake pump fitted to ex-TCCD No 45170, seen while the engine was at Storey Engineering's workshops at Hepscott, Northumberland, on 29 September 2013. It was stored there prior to its move to Bo'ness, where it is under restoration. It is to be named *Sir William McAlpine*. *Steve Frost*

The Steam Engines of World War II in Europe

One of the "austerity" shunting locomotives, of which a number is in use at W.D. depots, of the type recently ordered from various locomotive-building concerns by the Ministry of Supply. Some details of these engines are given on page 314.

British "Austerity" Saddle-Tank

THE Ministry of Supply has recently placed orders with locomotive manufacturing firms in this country for a number of 0-6-0 saddle-tank engines to a simple and robust design based by the Ministry on a standard shunter of a well-known locomotive building firm. The new engines are capable of shunting trains of 1,000 tons; also of dealing with military trains and mixed traffic generally for short journeys. Thus, though they are similar to a much used industrial type of engine, the locomotives are of generous proportions and powerful enough to rank with most shunting engines in British railway service. The principal particulars are as follows:—

Cylinders, dia.	18 in.
,, stroke	26 in.
Coupled wheels, dia.	4 ft. 3 in.
Evaporative heating surface, tubes	873 sq. ft.
,, ,, ,, firebox	87 sq. ft.
,, ,, ,, total	960 sq. ft.
Firegrate area	16.8 sq. ft.
Boiler pressure, per sq. in.	170 lb.
Tractive effort (at 85 per cent. boiler pressure)	23,870 lb.
Adhesion weight	48¼ tons
Weight of engine in working order	48½ tons
Water capacity of tank	1,200 gal.
Coal capacity of bunker	2¼ tons

The length of the engine over buffers is 30 ft. 4 in., the height from the rail to top of chimney is 12 ft. 1½ in., and the width overall is 8 ft. 2½ in. The wheelbase is 11 ft. A number of these locomotives is already in use, one of which is illustrated hauling a War Department train on page 263.

3.2 The Hunslet 'Austerity' 0-6-0 saddle tank engines

Construction and distribution

In 1942 the WD was looking for large numbers of shunting engines for use after the planned invasion of Europe. The Hunslet Engine Co of Leeds had already been building saddle tanks for industrial use for more than 70 years. It was not therefore surprising that the Ministry of Supply, having looked at and rejected the LMS 'Jinty' 0-6-0 tank, should look to Hunslet for its 'Austerity' design. The engines' construction should use as few resources as possible and they should have a life of at least two years. The basic design dated back to the 1920s, but had been improved significantly over the years. In 1941 Hunslet received an order for heavier engines from Stewarts & Lloyds of Corby. After only one engine had been delivered the order was cancelled and three of the remaining Class '50550' engines were purchased by the WD (Nos 65 to 67). These formed the basis of a new design modified to suit the WD's requirements. Details of the new engines were published in *The Railway Magazine* of September and October 1943

The WD Hunslet tanks: extracts from The Railway Magazine for September and October 1943. Reproduced with the permission of Mortons Media Group

(the magazine was then being published bimonthly due the paper shortage). By that October only 49 had been constructed, but many more were needed; further contracts were therefore signed with other manufacturers. Between 1943 and 1947, 377 engines were built as follows: Robert Stephenson & Hawthorn (90), Hudswell Clarkee (50), Bagnall (52), the Vulcan Foundry (50), Andrew Barclay (15) and Hunslet (120). A few of these engines were fitted with a Westinghouse brake pump for working with air-braked stock on the Continent.

Again, as with the 2-8-0s and 2-10-0s, not all of the engines were needed immediately by the WD. Many found work on the LMR and at other WD depots and docks around the country, but others were put into store at Longmoor. After D-Day some 90 engines were transported to the Continent between November 1944 and May 1945. Fifty of these were sent to Belgium where they worked on loan to SNCB; they were returned to the WD by the spring of 1946 and were put into store at Calais and Antwerp. Some of these engines were transferred to the Netherlands, which eventually had 27 of them. Although many were stored at Calais, they were not popular with the French who preferred the USATC 0-6-0 tanks, described in Chapter 4.1.

Post-war distribution

Although most of the 'Austerity' tanks were returned to Britain, the 27 left in the Netherlands were purchased by NS in 1947, Nos NS 8801 to 8827, but were all withdrawn or sold to industry by 1957. An earlier overseas transfer had occurred in the spring of 1946 when six engines were sold to Tunisian Railways (CFT), where they were numbered Nos 3.51 to 3.56; the last was not withdrawn until July 1971.

Those which remained on the Continent were gradually repatriated, the last in July 1947. In November 1945 WD No (7)1486 (RSH No 7295, 1945) was sent to the LNER at Doncaster for trials. These were successful and this engine, together with 74 others, was purchased by the railway. Most of these were unused, while some were still being built by Andrew Barclay; quite a bargain for the LNER. After modification for main-line use, all entered service as LNER Class 94. In 1948 60000 was added to their numbers when the LNER became part of BR. One of their last duties was on the Cromford & High Peak Railway in Derbyshire; among the last two engines to work there were Nos 68006 (ex-WD No 5094, HC No 1755, 1943) and 68012 (ex-WD No 5124, HE No 3174, 1944), which were not withdrawn until May and October 1967 respectively. Many more of the remaining WD engines were disposed of for industrial use at coal mines (now nationalised under the National Coal Board (NCB)), steelworks and docks, including the Port of London Authority and the Manchester Ship Canal. The coal Industry had been taking an interest in these engines for some time, but because Government orders took priority it was not until February 1944 that Hunslet was able to deliver one of them (HE No 3134, 1944) to Orgreave Colliery in Nottinghamshire. This was later to set a precedent for the NCB, which purchased a large number of the engines from the WD and ordered new ones from Hunslet.

Having disposed of its surplus but otherwise brand-new engines at knock-down prices, the MOD suddenly realised that it was short of locomotives! As a result, in 1951 14 more were ordered from Hunslet and delivered in 1952/53 (WD Nos 190 to 203 (HE Nos 3790 to 3803)). A number of the WD engines were given names at the depots where they became long-term residents. At the end of the war around ten 'Austerity' tanks were still based at Longmoor, but only three of these, now numbered and named 106 *Caen* (formerly *Spyck*), 118 *Brussels* and 157 *Constantine*, were still there in 1963, together with another WD engine, No 156 *Tobruk*. Both *Brussels* and *Tobruk* were oil-fired. By this time many of the saddle tanks were being withdrawn, including six – Nos 106/8/19/30/78 and 203 – that were sent to Barry scrapyard in 1963, where they were all broken up in 1965. In 1968, under what was now the Ministry of Defence (MOD), few 0-6-0STs were still in service. Numbers were therefore rationalised. Nos 190 to 194 and 198 became WD Nos 90 to 94 and 98, while Nos 200 to 202 became Nos 95 to 97. Most of the MOD's 'Austerity' tanks were withdrawn and scrapped in 1970/71. Three engines, Nos 92 *Waggoner*, 197 *Sapper* and 98 *Royal Engineer*, were, however, retained, in immaculate condition, for special workings (see the accompanying pictures). A further engine, No 196 *Errol Lonsdale*, the last of its class to work at Longmoor, and *Sapper* had retained their old numbers as they were scheduled for preservation by the MOD – see below.

In 1961 Hunslet carried out a number of experimental improvements to one of the NCB's engines, No 2876 of 1943 (ex-WD (7)5027). These involved fitting a gas producer system and mechanical stoker, and a modified grate, brick arch and blastpipe. No 2876 then worked at several collieries in Yorkshire until scrapped in 1976. In April 1963 Hunslet purchased another engine, WD No 168 (ex-(7)5019, HE No 2868, 1943) and rebuilt it in the same way. The engine became HE No 3883 and was sent to BR's Swindon Works for testing. At trials around Oxford it became the last steam engine ever to be tested using the ex-GWR's dynamometer car. No 2868 was then sold to the NCB, working in Yorkshire until withdrawn in 1976. It was subsequently preserved. The measures were found to be successful and the features were incorporated into subsequent new engines, while others were rebuilt with them. Seventeen of the WD saddle tanks were similarly modified; most were Hunslet engines, but two of Bagnall's engines were rebuilt at that company's own works. Other NCB engines were fitted with Giesl ejectors.

In all, Hunslet built almost 80 engines for the NCB, the last two, Nos 3889 and 3890, in 1964. In addition, two engines had been built for the NCB by Robert Stephenson & Hawthorn in 1955 (RSH

Nos 7751/2). Hunslet's No 3890 became the last British steam engine built for the home market in the industrial age. In all, some 485 saddle tanks were built by the various firms from 1943, and many of the MOD and industrial engines were still at work long after steam had finished on BR in August 1968.

Preservation in Europe and Tunisia
The 'Austerity' saddle tanks are by far the most numerous class of preserved steam engine in the UK. According to the Wikipedia 'List of preserved Hunslet Austerity 0-6-0ST Locomotives' (which includes those by other manufacturers) some 72 examples have been preserved. Thirty-nine of these are ex-WD engines, including nine of the 14 engines built new in 1952/53. In the event it was *Waggoner* and *Royal Engineer* that were preserved by the MOD, not *Sapper* or *Errol Lonsdale*, which were purchased privately. The surviving WD/MOD engines are listed in Table 4, including those preserved abroad.

Table 4a includes a selection of those built after the war for the NCB and for other industrial use.

'Austerity' saddle tanks working for the WD/MOD

The WD (MOD from 1964) retained a large stock of steam engines after the war. This included a group of 'Austerity' saddle tanks at the Longmoor Military Railway (LMR), which were used both for shunting and training purposes. One example was 0-6-0ST WD No (7)5275 (No 177 after 1952) *Matruh* (RSH No 7205, 1945), which is seen at the LMR depot on 2 September 1950. The engine was scrapped in 1959.
Frank Hornby

One of the last of the active 'Austerity' saddle tanks on the LMR in the late 1960s was MOD No 196 *Errol Lonsdale* (HE No 3796), one of 14 new 'Austerity' saddle tanks ordered from Hunslet in 1952/53. It is seen here on the LMR in October 1969 crossing the road at Oakhanger with a passenger train. In October 1965, together with another MOD 'Austerity' tank, it took the part of BR Class 94 No 68011 during the filming of *The Great St Trinian's Train Robbery* at the LMR. The engine was subsequently preserved (see page 80).
M. E. J. Deane collection, courtesy of Ian Bennett

The MOD's Bicester Military Railway opened in 1942 to serve the WD's Central Ordnance Depot. Among the other 'Austerity' saddle tanks remaining with the MOD after the end of steam on BR, specifically to work special trains, was No 197 *Sapper* (HE No 3797), another of the 14 new engines ordered in 1952/53. On 16 October 1971 a railtour, the 'Western Trooper', was organised by the Southern Locomotive Preservation Co to visit two MOD depots where steam was still in use. The train left Waterloo behind 'Western' diesel-hydraulic Class 52 No (D)1033 *Western Trooper* and travelled via Ascot, Reading and Oxford to Bicester (London Road). Here *Sapper* took the train on a tour of the Bicester depot, and is shown at the head of the railtour in the depot's exchange sidings with BR. The engine is now at the K&ESR as its No 25 *Northiam*. *Ray Ruffell, Slip Coach Archive*

After visiting Bicester the 'Western Trooper' railtour ran to Long Marston via Banbury. The railway there had opened in 1940 to serve the WD Central Ordnance Depot, and there MOD 'Austerity' saddle tank No 98 *Royal Engineer*, another engine retained for special workings, hauled a two-coach special train over the railway. No 98 was also ordered from Hunslet in 1952/53 (No 3798) and ran as WD No 198 until 1968 when it became No 98. It was the last MOD 'Austerity' to be withdrawn in 1991 and was subsequently preserved on the Isle of Wight. Long Marston Depot was privatised in 2004 and is now used to store rolling stock owned by the various train leasing companies.
Ray Ruffell, Slip Coach Archive

The third 'Austerity' saddle tank retained for special workings was No 192 *Waggoner* (HE No 3792), yet another of the 1952/53 engines. The engine, now numbered 92, is seen at the WD/MOD Marchwood Depot in Hampshire with a short train of ex-BR Mark 1 coaches during an open day on 22 July 1978. The depot is reached over a spur from the Fawley branch and remains open today. Like No 198, the engine was initially preserved at the Museum of Army Transport at Beverley in East Yorkshire, but is now on the Isle of Wight. *M. E. J. Deane collection, courtesy of Ian Bennett*

Disposed of by the WD at Barry scrapyard

Above: In 1963 six ex-WD 'Austerity' saddle tanks were sent to Woodham's scrapyard at Barry for breaking up. One of them, No 106 (Hunslet No 2889, 1943) is seen there on 23 June 1963. The engine was once named *Spyck* and was numbered (7)5040 until 1952. During the 1940s and '50s it spent long spells at the Marchwood Military Depot together with periods on loan to the Port of London Authority. By 1960 it was at the LMR, but was withdrawn in December 1961. All the Barry engines were cut up by March 1965.

Ex-WD/MOD 'Austerity' saddle tanks preserved in the UK and Europe

Above right: Ex-WD No 200 (HE No 3800, 1953) was among the last of the 14 'Austerity' saddle tanks built in 1953. In 1968 it was renumbered MOD No 95 and went to the MOD Shoeburyness Depot, but was put into store there. In 1971 it was sold to the K&ESR where it was numbered 24 and named *William H. Austen*. The engine is seen taking water at Tenterden on the K&ESR on 25 June 1978. It went to the Colne Valley Railway in 2014 and is currently under restoration there.

Right: Ex-WD No 192 *Waggoner* (HE No 3792, 1953) is seen again now in preservation leaving Havenstreet station on the Isle of Wight with a train from Smallbrook Junction to Wootton on 25 September 2015. After the Museum of Army Transport at Beverley closed in August 2003 the engine went on loan to the Isle of Wight, where it was returned to steam. The engine is running again, still in its LMR Prussian blue livery.

Ex-LMR 'Austerity' tank WD No (7)1505 *Brussels* (Hudswell Clarke 1782, 1945) has been preserved on the K&WVR for many years, and is seen in the museum at Oxenhope on 8 July 2012. It went to the Continent in February 1945 together with nine other 'Austerity' tanks. There it was one of 50 0-6-0STs loaned to SNCB and allocated to Merelbeke shed. It returned to Britain and was converted to an oil-burner at Longmoor after being involved in an accident there in 1953. By then the engine was numbered WD118.

WD No (7)5105 (HE No 3155, 1944) was one of ten 'Austerity' saddle tanks sent to the Continent in February 1945. After the war it remained in the Netherlands where in 1947 it was sold to Netherlands Railways (NS), working as its No 8815. On its retirement it was sold to the Laura Colliery at Eygelshoven, but in 1973 it went to a Dutch scrap merchant at Tilburg. It was subsequently purchased for preservation and returned to Britain for restoration in 1988, initially at the Northampton Steam Railway. It is seen here on 14 March 2015 while working on the Spa Valley Railway at Eridge with a train for Tunbridge Wells West. The engine is carrying WD green livery and is named *Walkden*. It is now at the Ribble Steam Railway near Preston.

Right: WD tank engine No (7)5115 (HE No 3165, 1944) was also among the ten 'Austerity' saddle tanks sent to the Continent in February 1945, and like No (7)5105 it remained in the Netherlands, later to work on NS as No 8826. On its retirement it was sold again, this time to the Julia Colliery at Eygelshoven, where it was named *Julia V.* In 1975 it went to the Tilburg scrapyard, but is now preserved by the South Limburg Steam Train Company. The engine is seen partly dismantled after arrival at the museum in June 2000, where it is still under restoration.
Courtesy of the South Limburg Steam Train Company

Above right: Ex-WD 'Austerity' saddle tank No 196 *Errol Lonsdale* (HE No 3796, 1953), seen earlier at Longmoor, remained there until the facility closed in October 1969. It was preserved first on the K&ESR before moving to the Mid-Hants Railway, then to the SDR, where it was pictured at Buckfastleigh on 14 October 2008. Although still carrying its WD nameplate, it had been returned to the guise it took during the making of *The Great St Trinian's Train Robbery*. The engine is now at the Stoomcentrum Maldegem, Belgium.

Ex-WD 'Austerity' saddle tanks working for BR

Right: Seventy-five of the WD's 'Austerity' saddle tanks were bought by the LNER in 1946, becoming LNER/BR Class J94. One of them, No 68046 (WD No (7)5331 (Vulcan Foundry No 5321, 1945)) is seen out of use at Darlington Shed (51A) on 1 August 1964. Built too late to be taken to the Continent, it was one of dozens of new engines put into store at Longmoor from August 1945. Many of the LNER engines came from this store. No 68046 initially became a Neasden (34E) engine before going to York (50A) and then Darlington (51A). It was withdrawn in September 1964 and broken up at Cleveland Dockyard, Middlesbrough, later that month.

Below: Two of the ex-BR Class J94 'Austerity' saddle tanks, Nos 68077 and 68078 (WD Nos (7)1466 and (7)1463), have been preserved. Although built for the WD by Andrew Barclay (Nos 2215 and 2212 in 1946 and 1947 respectively), they were both delivered new to the LNER. Then numbered 8077, No 68077 was initially allocated to Immingham shed, then after spells at Hornsey and Boston it was withdrawn from Colwick shed in December 1962. The engine was then sold to the NCB and worked initially at Orgreave Colliery, Yorkshire. It was out of use at Maltby Main Colliery by May 1970, and in 1971 was sold to the K&WVR, where No 68077 is seen in the stock shed at Haworth in September 1989. It is now preserved on the Spa Valley Railway in Kent. No 68078 is at Hope Farm, Sellindge, also in Kent.

Above: Ex-WD 'Austerity' saddle tank No (7)5124 (HE 3174, 1944) was delivered to the WD Depot at Burton Dassett, Warwickshire, in 1944 and a year later was fitted with experimental radio control equipment at Kineton on the Melbourne Military Railway. It was one of the 75 engines purchased by the LNER in May 1946, becoming its Class J94 No 8012. It received its BR number, 68012, in May 1951. When seen at Rowsley shed (16J), Derbyshire, on 20 May 1962 it had recently moved from Gorton shed (9G), Manchester. At Rowsley it worked on the Cromford & High Peak Railway and, when pictured, was fitted with the oval buffers used on that line to avoid buffer-locking on the sharp curves. The engine was the last of its class to be withdrawn, in October 1967, but was then briefly loaned to the NCB before being broken up less than a year later. *Frank Hornby*

Ex-WD 'Austerity' saddle tanks working for the National Coal Board
Above: One of the 'Austerity' tanks at Gresford Colliery, Wrexham, was RSH No 7135, of 1944. It was built for the WD as No (7)5185, and was one of the last 24 engines sent to the Continent in May 1945 and put into store at Calais. By 1947 it had returned to Britain and was sold to the NCB, going to Llay Main Colliery, Denbighshire. In the late 1960s it went to Gresford Colliery and is seen in steam there on 22 February 1970. The engine ended its days at Bickershaw Colliery, Lancashire, in 1978. In 1981 its boiler, cylinders and other parts were used in the NRM's replica of the GWR's broad-gauge engine *Iron Duke*. *Barry Payne*

The Steam Engines of World War II in Europe

Above: NCB 'Austerity' tank No 8 of Mountain Ash Colliery, South Wales, stands outside the loco shed there on 6 March 1970. Although in steam, the instructions on its smokebox door read 'Do Not Move'. Built for the WD as No (7)5189 in 1944 by RSH (No 7139), the engine was used initially at the Command Ordnance Supply Depot, Didcot. It later went to Longmoor where it was named *Rennes* and given a Prussian blue livery, becoming WD No 152. In 1960 Hunslet purchased the engine and rebuilt it (as HE No 3880) before selling it back to the NCB. Mountain Ash Colliery closed in September 1979 and, in 1981, the engine was purchased for preservation. It was initially on display at the Big Pit Mining Museum, but is now at the Dean Forest Railway. *Barry Payne*

Above and left: NCB 'Austerity' tank No S134 was built for the WD by HE (No 3168) in 1944. At first it was numbered WD (7)5118, but became No 134 after 1952. It went to the Bicester Depot in 1951, where it carried the name *King Feisal* (of Iraq). In 1965 it was sold to the NCB and in 1966 was based at Primrose Hill Colliery, Swillington, Leeds, where it is seen at work early in 1970 shortly before the colliery closed. The engine was fitted with a mechanical stoker and a gas producer system, and was used for tests by Hunslet in 1981; it was then transferred to Wheldale Colliery at Castleford. It was last reported working there in July 1982, when it is thought to have been the last active NCB steam engine in Yorkshire. In November 1982 it was purchased for preservation by the Embsay & Bolton Abbey Steam Railway and named *Wheldale*. It worked there for a while, but was then put into store where it remained for many years. It is currently being overhauled. *Steve Roberts*

Left: Littleton Colliery's 'Austerity' tank NCB No 7 (HC 1752, 1943) was built for the WD (No (7)5091) and initially worked at the Marchwood Military Railway. It spent time on loan to the SR at Southampton Docks, then worked at other WD depots until sold to the NCB in 1950. The engine is seen heading for the exchange sidings at Penkridge, on the BR line between Wolverhampton and Stoke-on-Trent in Staffordshire, with a train of loaded coal wagons from Littleton Colliery, Cannock, on 22 February 1970. In 1978 the engine was transferred to NCB Bold Colliery near Liverpool, as seen in the following photo. *Barry Payne*

Above: Littleton Colliery's No 7 is seen again after transfer to NCB Bold Colliery, Liverpool, where it was named *Robert*. While there it took part in the Grand Parade of engines at Rainhill in May 1980, carrying its NCB Bold Colliery livery of lined-out green. After withdrawal it was displayed at the Chatterley Whitfield Mining Museum in Staffordshire. After storage at various locations, in 2007 it went to the Great Central Railway, Loughborough, where it was restored to steam (see the following photo).

Above: A number of the preserved 'Austerity' saddle tanks have been restored in the guise of one of BR's Class J94s, and one of these is ex-NCB *Robert* (see the previous photos). It is seen at Loughborough on the Great Central Railway as No 68067 on 12 November 2017.

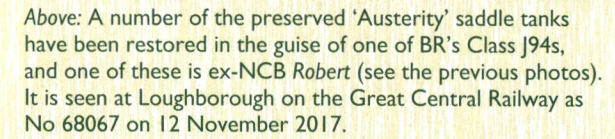

Above centre: WD No (7)1515 was one of the 'Austerity' tanks built by RSH (No 7169, 1944). It was loaned to the MoF&P in 1944 and worked at the Ashington Opencast Coal site until transferred to the NCB (Opencast) in 1952. By 1967 it was working at the Swalwell Opencast site at Whickham in County Durham, but in 1973 the engine was sold to the East Somerset Railway at Cranmore, Somerset. It is now at the Pontypool & Blaenavon Railway in South Wales, where it is seen on 27 May 2013 carrying the name *MECH. NAVVIES LTD*.

Above right: In April 1963 Hunslet purchased one of the MOD's engines, No 168 (ex-WD (7)5019, HE No 2868, 1943) and fitted it with a gas producer system, a mechanical stoker, and modified grate, brick arch and blast pipe. The rebuilt engine, HE No 3883, 1963, was sent to BR's Swindon Works for testing. At trials around Oxford it became the last engine ever to be tested using the ex-GWR's dynamometer car. No 3883 was then sold to the NCB, working in Yorkshire until withdrawn in 1976. It was preserved at 'Rocks by Rail' in Rutland, where it is seen, unrestored, on 15 June 2008. It subsequently moved several times and is now at the Midland Railway Centre, Butterly.

Right: WD No (7)5008 (HE No 2857, 1943) was among the first 18 'Austerity' tanks sent to the Continent in November 1944. For a time it was based at the SNCB shed at Antwerp Dam, but was repatriated in 1946. In April 1947 it went to the Old Silkstone Colliery in Yorkshire under the auspices of the MoF&P before transfer to the NCB. It was fitted with an underfeed stoker by Hunslet in 1965 while based at Dodworth Colliery, Yorkshire. It worked finally at the Cadley Hill Colliery in Derbyshire before going to the Lavender Line at Isfield in East Sussex. and being named *Swiftsure*. Now at the Nene Valley Railway, the engine, running in khaki livery and carrying its ex-WD number 75008, is seen after crossing the River Nene at Wansford with a train for Peterborough on 29 May 2016.

Post-war Hunslet saddle tanks built and working for the NCB and other industries

Right and below: One of the Hunslet tanks built new for the NCB was No S116 (HE No 3836, 1955). It initially went to Waterloo Main Colliery, Yorkshire, where it was named *Sylvia*. The engine was rebuilt by Hunslet in 1963 with an underfeed stoker and a gas producer system, returning to Waterloo Colliery before going to Primrose Hill Colliery Swillington, Leeds, in 1964. The engine, now unnamed, is seen at work there on 22 February 1970. Primrose Hill Colliery closed in March 1970 and, after a time at the NCB's Allerton Bywater Central Workshops, No S116 was broken up in 1973.
Barry Payne

Above: The heritage Gwili Railway is based at the former BR station at Bronwydd Arms, north of Carmarthen in West Wales. It is home to two ex-WD 'Austerity' saddle tanks. One of them, ex-WD No (7)1516 (RSH No 7170, 1944), was loaned to the MoF&P in 1944 and worked at various collieries in north-east England. In 1959 it was transferred to South Wales, ending its days at the Graig Merthyr Colliery at Pontarddulais until moving to the Gwili Railway in 1980. There it is reported to have been rebuilt with parts from sister engine W. G. Bagnall No 2758, 1944 (ex-WD (7)5170). The engine is pictured on the Gwili Railway at Bronwydd Arms in May 1994 carrying the name *Welsh Guardsman*.

Right: Another NCB engine purchased direct from Hunslet was HE No 3825 in1954. It spent its working life in the Kent coalfield, first at Betteshanger as NCB No 9, then in 1972 at Snowdown Colliery, where it was pictured in 1979. Approaching is ex-BR diesel shunter No 12131, which was withdrawn by BR in March 1969. No 12131 was one of more than 130 diesel shunters built for the LMS/BR at Derby and Darlington Works, using English Electric engines, and were the precursors of BR's Class 08 diesel shunters. In 1981 the Hunslet engine was preserved at what was then the Main Line Steam Trust at Loughborough. It subsequently moved to several other heritage railways and is now based on the Stainmore Railway at Kirkby Stephen East station in Cumbria, where it is running in the guise of BR J94 No 68009. The diesel shunter it is now preserved on the North Norfolk Railway.

Left: Hunslet No 3694 was one of ten engines built for the NCB in 1950. It initially worked at Haydock and other Lancashire collieries until moving to Bold Colliery in 1962, where it was named *Whiston*. Here it was a stablemate to ex-WD No (7)5113 (Hunslet No 3163, 1944) and *Robert* (ex-WD No (7)5091,(HC No 1752, 1943))(see previous photographs). It is now preserved on the Foxfield Railway, where it was seen on 17 July 2010 sporting its former Bold Colliery livery.

Right: A further ten saddle tanks were built for the NCB in 1952, and one of them, Hunslet No 3777, worked initially at Baggeridge Colliery in Staffordshire. It moved to other Staffordshire collieries in 1968, ending its days at Wolstanton Colliery. In 1977 it went to the Churnet Valley Railway (CVR) for preservation, and was converted into an ex-BR Class 94, numbered 68030 and named *Josiah Wedgwood*. The engine is seen at Weybourne on the North Norfolk Railway while working a train from Sheringham to Holt on 3 September 2011. After leaving the CVR it spent time at several other heritage railways, and is now at Llangollen. The second engine at Weybourne is ex-GWR pannier tank No 7717; this later became London Transport No L99 and was carrying the LT maroon livery here.

Above: Nine more saddle tanks were built for the NCB in 1956, one of which, Hunslet No 3839, was initially based at the NCB's Rawnsley loco sheds, Staffordshire. It later worked at the Cannock collieries. In 1973 it was preserved at the Foxfield Railway (FR), where it was numbered 7 and named *Wimblebury*. It is seen carrying its NCB livery on the FR on 17 July 2010.

Left: In addition to the 'Austerity' saddle tanks built new for the NCB, other industries were also catered for. These included Stewarts & Lloyds Minerals (later British Steel) for work at Corby steelworks in Northamptonshire together with its associated ironstone railways. One of them, named *Juno* (Hunslet No 3850, 1958), worked at the company's Buckminster Quarries on the South Lincolnshire/Leicestershire border. When the quarries closed *Juno* was preserved, first at Quainton Road, Buckinghamshire, then on the Isle of Wight Steam Railway. It has now been cosmetically restored and displayed at the NRM's site at Shildon in County Durham, where it was pictured in 2012. *Steve Frost*

3.3 The WD 'Austerity' 2-8-0s

Construction and distribution

The Standard WD 'Austerity' 2-8-0 was designed to meet the expected need for heavy goods engines to move troops and supplies into Europe after an invasion. The resources needed to build them were also minimised. R.A. Riddles was put in charge of both their design and production in consultation with the North British Locomotive Co (NB) and the Vulcan Foundry (VF). The engines were essentially a stripped-down version of the Stanier 8Fs. They were described in detail in *The Railway Magazine* of January and February 1943 and a photograph of the first engine appeared in the July and August issue (see the accompanying illustrations). As neither the solid-cast balance weights attached to the driving wheels nor their springing could be adjusted after assembly, they were always considered to be rough-riding engines, and once in Europe it was thought that the state of the track would rule out any fast running. The engines were nevertheless expected to haul 1,000-ton trains at up to 40mph. Some engines were fitted with train-heating and their tenders with water scoops, while their design allowed them to be easily converted to

Left, and above: Extracts from *The Railway Magazine* for January and February 1943: 'New Locomotives for War Work'. *Reproduced with the permission of Mortons Media Group*

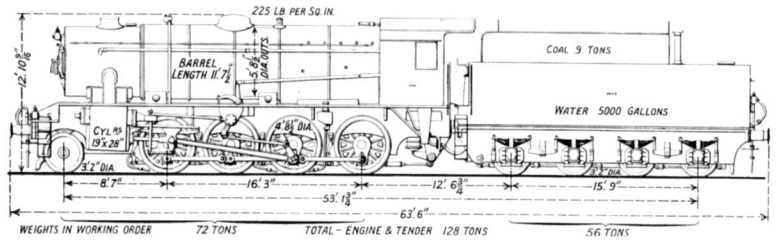

The first British "austerity" 2-8-0 locomotive at Glasgow before its trial run.
This locomotive was described in the Jan.-Feb. RAILWAY MAGAZINE

An extract from The Railway Magazine for July and August 1943: 'New Locomotive for War Work'. Reproduced with the permission of Mortons Media Group

'Austerity' saddle tanks, a number of engines were used by the Belgian SNCB or Netherlands Railways NS. One hundred and seventy engines worked in Belgium up to the end of 1945, but were then returned to Britain or went to NS. The NS fleet finally numbered 237 engines, Nos 4301 to 4537. Three hundred and seventy-three engines were temporarily loaned to the SNCF.

Post-war distribution
In 1945 the WD informed the REC that some of the engines were surplus to requirements, although the number on offer was later reduced. The first eight arrived back in Dover in November 1945, followed by 470 more by the spring of 1946. Many of the returning engines were found to need extensive maintenance and were put into store in various places. In November 1946 the LNER purchased 200 WD 2-8-0s (it already had 190 of them on loan) and classified them '07'. It undertook a number of modifications, including removing the air compressors and fitting glass to the windows. The LNER was still interested in acquiring more of the engines and by December 1947 a further 270 were on loan to the company. This had also allowed a number of the LNER's Stanier engines to be returned to the LMS.

After nationalisation in January 1948 the class was tested during the Locomotive Exchanges of that year. Although found to be less powerful than comparative 2-8-0s, they had the advantage as far as coal and water consumption was concerned. As a result Mr Riddles was instructed to negotiate the purchase of the 425 engines that were already on loan to the LNER (270), GWR (89), SR (50) and BR itself (16), together with 108 more currently in store. A deal was concluded, with each of the engines costing on average £2,688. With the 200 LNER-owned engines, BR ended up with a fleet of 733; they were to be classified '8F WD Class', and were not to be referred

oil-burners. The engines were expected to last at least two years running under wartime conditions!

In all, 935 WD 2-8-0s were ordered, 345 from NB and 390 from VF. Of those ordered from NB, 130 replaced an existing order for Stanier 8Fs. The first NB engine was completed in January 1943, while the first VF example appeared some two months later. The whole order was complete by May 1945, when the last engine left the Vulcan Foundry (VF No 5255), named *Vulcan*, a name it was to carry during its BR days as No 90732. As the invasion of Europe, D-Day, did not happen until June 1944, 450 of the earlier engines were lent to British railway companies, 350 to the LNER and 50 each to the LMS and SR. The rest were either used by the WD or put into store.

The USATC Class S160 2-8-0s (Chapter 4.2) also worked on the national network when they first arrived in Britain, many of them on the GWR. These engines were sent to Europe before the WD 2-8-0s, so the GWR received 89 WDs in compensation. All but three of the WD engines went to France and Belgium in batches from September 1944; the three that remained were used by the WD at home, and included No (7)9250 (VF No 5193, 1945), which went to the LMR and remained there for many years. In Europe the class was deployed widely in France, Belgium, the Netherlands and in Germany. During the first half of 1945 222 were lent to the USATC, one of which, WD No (7)8678, was scrapped after being severely damaged in an accident. As with the

WD 'Austerity' 2-8-0s working for BR

Above: Nine hundred and thirty-four 'Austerity' 2-8-0s were built by North British and the Vulcan Foundry for the WD from 1943. WD 2-8-0 No (7)7451 (VF No 4967, 1943), seen here, was one of 450 engines loaned to the main-line railways until required in Europe. It worked initially on the LNER, then the GWR, before being recalled and sent to France in January 1945, returning home early in 1947. It was then one of 42 of its class stored at Kingham in Oxfordshire, where it was pictured on 27 September 1949. The engine unusually has its WD number displayed on its smokebox. Seven hundred and thirty-two of the engines later became BR Class WD, No (7)7451 becoming BR No 90595, receiving its new number in March 1951. It was withdrawn from Lancaster shed (24J/10J) in February 1964. *Frank Hornby*

Above right: WD 2-8-0 No (7)7480 (VF No 49967, 1943) also worked for the LNER before being sent to the Continent. It returned in April 1946 and was put into store at Feltham. It later became one of 25 engines to work on BR(SR) and is seen here heading a freight in Feltham Yard on 26 March 1949. It received its BR number, 90604, in May 1950. All the SR's engines went to other BR regions in 1950/51, No 90604 going to BR(LMR). It was withdrawn from Wakefield shed (56A) in what was by then BR(NER) in December 1963. *Frank Hornby*

Above: WD 2-8-0 No (7)8580 (NB No 25341, 1944) was initially put into store at the LMR before going to the Continent, where it was one of 191 engines lent to the USATC in the spring of 1945. It was one of the last to return home in 1948 and was stored at Stratford shed as BR(ER) No 63107, where it was pictured on 8 March 1948. It received its BR number, 90379, in September 1949, and was also withdrawn from Wakefield shed (56A), in March 1966. *Frank Hornby*

to as 'Austerities'! Their numbering originally followed the BR(ER) model from, Nos 63000 to 63199, but this was soon replaced by numbers familiar to most enthusiasts, Nos 90000 to 90732.

The greatest number could be found working on BR's Eastern and North Eastern Regions, while some ran on every BR Region. BR(SR) relinquished its WDs to the LMR in 1950/51, while those on BR(WR) went to either the LMR or ER in 1962. During their time with BR various measures were taken to improve them, including modifications to their boilers and a reduction in the diameter of the blastpipe orifice. Nothing could be done, however, to reduce their rough riding. The first engine to be withdrawn, from BR(ER)'s March depot (31B), was No 90083 (NB No 8513, 1944) in December 1959. However, the last 24 were not withdrawn from their BR(NER) sheds until September 1967, having outlasted their predicted two-year life expectancy by 43 years! None of the BR engines was preserved.

Two of the WD 2-8-0s, WD Nos (7)7337 *Sir Guy Williams* (NB No 25205, 1943) and (7)9250 *Major General McMullen* (VF No 5193, 1945), remained at the LMR after the war, and were renumbered 400 and 401 in 1952. No 401 was the first to be scrapped in 1958, while No 400 remained intact until 1967. In addition it is thought that one of the engines, WD No 9189 (VF No 5132, 1944), was exchanged for a USATC Class S160 so that comparisons could be made. For completeness, it should also be noted that 12 of the WD engines went from storage at the LMR to Hong Kong in 1946/47, and did not return.

The 237 engines working on NS remained in service until replaced by progressing electrification in the early 1950s, the last four being withdrawn in 1954. Two NS engines, Nos 4383 and 4464 (WD Nos (7)8528 and (7)9257), were sold to Sweden in June 1953, where they became SJ Class G11 Nos 1930 and 1931.

Right: No 90516 (ex-WD No (7)9253 (VF No 5196, 1945)), was one of the 200 engines sold to the LNER as Class O7. For a short time it became BR No 63195 before receiving its final BR number in December 1949, No 90516. It was withdrawn in November 1965. When seen at St. Philip's Marsh shed (82B), Bristol, on 8 March 1963 the engine was allocated to the former GCR shed at Woodford Halse (2F).

Be;ow right: Many of the 'Austerity' 2-8-0s worked in north-east England. Here No 90452 (ex-WD No (7)8634 (VF No 5035, 1945)) drifts past the signal box and over the busy Anlaby Road level crossing in Hull while running light engine to Dairycoates shed on 30 July 1964. At the time the construction of a replacement road flyover over the Anlaby Road was well under way. The engine was one of 222 lent to the USATC in the spring of 1945 for use in France, Belgium and the Netherlands. All were returned to the WD in August 1945, and after a period in store the engine re-crossed the Channel in December 1945. It was then sold to the LNER, where it was numbered 3130. For a short time it became BR No 63130 before receiving its BR number in April 1949. It was withdrawn in June 1965.

Preservation

All 733 BR examples were broken up, but, as noted in Table 4, one of the WD engines – one of the pair that went to the Sweden in 1953 – ex-WD No (7)9257 (VF No 5200, 1945), is now preserved in Britain. It was put into store in 1958 but was later purchased for preservation and arrived at the K&WVR at Haworth in January 1973. Much work was needed to convert it back to a BR engine, including building a new tender – the original had been reduced in length in Sweden to enable the engine to fit on SJ's turntables. The engine can now be seen in action again on the K&WVR, carrying the fictitious BR number 90733.

Above: WD No 90254 (ex-WD No (7)7355 (NB No 25230, 1943)) is seen dumped at its home shed of West Hartlepool (51C) in September 1967. It had been withdrawn on 31 January 1967 and was broken up later in September. The last of the class were withdrawn during September 1967. In 1943 No 90254 was one of many engines that went to the main-line railways prior to being sent to the Continent, being used by the LNER based at Tyne Dock, then Heaton. The engine was shipped to France in January 1945, but by March 1946 it was back and based at the SR shed at Bricklayers Arms. After nationalisation all BR(SR)-based WDs went elsewhere in 1950/51. This example eventually returned to its wartime stamping grounds in north-east England. *Steve Frost*

Below: The last of the WD 'Austerity' 2-8-0s, No (7)9312 (VF No 5255, 1945), was named *Vulcan*. It was sent to France during 1945 but was immediately put into store at Calais; it then returned home during 1947 and in November was loaned to the LNER. It was later purchased by BR, becoming No 90732 in May 1950. The engine is seen at Doncaster shed (36A) during the early 1960s. It was withdrawn from Frodingham shed (36C) in September 1962. *Colour Rail*

Above: WD 'Austerity' 2-8-0 No 90711 (ex-WD No (7)9273 (VF No 5216, 1945)) is also seen in Hull as it approaches the new Hessle Flyover with a train of coal empties on 27 July 1964. Before its purchase by BR it was stored for a time at Marston sidings near Swindon. It became BR No 90711 in June 1950 and was withdrawn in January 1967.

WD 'Austerity' 2-8-0 sent to Holland and Sweden and now preserved in the UK

Left: None of the 'Austerity' 2-8-0s that worked for BR survived into preservation. WD No (7)9257 (VF No 5200, 1945) was one of the relatively few engines of its class that went to the Continent during 1945, but did not return home. It soon found itself in the Netherlands where it was one of 75 engines purchased by Nederlandse Spoorwegen and numbered NS No 4464. In June 1953 it was sold to Sweden as SJ Class G11 No 1931. There it worked until retirement, when it became part of the Swedish Strategic Reserve. It was later purchased for preservation and brought to the K&WVR, where it is seen soon after arrival at Haworth in January 1976.

Bottom left: SJ No 1931 is seen again in April 1976 drifting downhill with a train from Oxenhope to Keighley. It is coupled to another former WD engine, ex-Longmoor Military Railway 'Austerity' tank WD No (7)1505 *Brussels* (Hudswell Clarkee 1782, 1945). The engine spent some time in Belgium in 1945 and is pictured more recently in the museum at Oxenhope on page 69. *Steve Frost*

Below: After an extensive rebuild the 'Austerity' 2-8-0 appeared as BR Class WD No 90733 (the last BR number was No 90732, named *Vulcan*). The rebuilt engine leaves Keighley with a train for Oxenhope on 11 October 2014. *Steve Frost*

The rebuilt engine is seen at Loughborough during a visit to the heritage Great Central Railway on 11 October 2009 with its demonstration mineral train, a replica of the 'Windcutter' coal trains which ran from the Nottingham coalfields to Woodford Halse during the 1960s.

WD 'Austerity' 2-8-0 No (7)9257, in the guise of BR Class WD No 90733, is seen again at Loughborough on the heritage Great Central Railway on 11 October 2009.

3.4 The WD 'Austerity' 2-10-0s

Construction and distribution

Although ten-coupled engines had been used in Europe since the turn of the 20th century, the wheel arrangement was never popular in Britain. The 150 Standard or 'Austerity' WD 2-10-0s were the first ten-coupled engines to be built in Britain for home consmption in any number greater than one! In contrast, as was described in Chapter 1, the Germans built around 10,000 during the Second World War. Like the WD 2-8-0s, the class was designed by R. A. Riddles and shared many of their features and, like them, needed the minimum of resources to manufacture. The 2-10-0s weighed 78 tons 6cwt and had a tractive effort of 34,215lb. They had a lighter axle load than the 2-8-0s, a larger boiler and a rocking grate like the USATC Class S160s, described in Chapter 4.2. The central driving wheels were flangeless, like the later BR 9F 2-10-0s. The innovative design of the firebox allowed the engines to steam more freely than their 2-8-0 cousins. The tenders of both classes were the same, while, the 2-10-0s, were relatively easy to convert to oil-firing.

The order for all 150 engines went to the North British Locomotive Co (NB), which switched production from the 2-8-0s. One hundred engines were ordered in June 1943, the other 50 in March 1944. The first were delivered from December 1943 and, like the 2-8-0s, were at first offered on loan to British railway companies. Here they were regarded with some suspicion by the respective railway CMEs, due to their long wheelbase. The CME of the LMS, one Sir William Stanier, said that the engines would not be allowed on his network until certain trials, involving points alongside station platforms, were carried out. These tests were soon completed successfully at the LMR. When the first engine went to St Rollox shed in Glasgow it was able

Photo] *[C. R. Gordon Stuart*

British 2-10-0 austerity locomotive No. 3692 on an empty coal train near Tring, L.M.S.R

An extract from *The Railway Magazine* for September and October 1943: 'British 2-10-0 austerity on the LMSR'. The engine, carrying WD number 3692 (NB No 25478, 1943), was then based at Rugby shed. It went to the Continent in October 1944 and later worked for the Netherlands Railways (NS) as its No 5021. *Reproduced with the permission of Mortons Media Group*

'Austerity' 2-10-0s working for the WD/MOD

The first of the North British built 'Austerity' 2-10-0s to arrive at the LMR in March 1944 was WD No (7)3651 (NB No 25437, 1943)) which later became WD No 600 *Gordon*. It was later joined by WD No (7)3797 (NB No 25643, 1945)) *Sapper*. The engine later became WD No 601 *Kitchener*. Both engines remained at Longmoor for many years, *Gordon* until the railway closed. The latter is seen at Longmoor on 2 September 1950 still numbered WD 73651. *Frank Hornby*

to demonstrate its flexibility on sharp curves. Thus the prejudice against ten-coupled engines in Britain was (partly) dispelled. In all 51 2-10-0s went new to Glasgow's LNER Eastfield shed for running in, the last in March 1944.

By this time the earlier engines were being dispatched to the Middle East, while one, WD No (7)3651, went to the LMR where it became long-term resident *Gordon*. Many others went south and worked from the sheds at Stratford and New England, Peterborough. Engines were also delivered new to the LMS, and the LNER engines were gradually transferred to that company. In November 1944 all 79 engines by then on loan to the LMS were recalled and, after servicing at the LMR, were sent to the Continent.

In March 1945 the next batch of 50 went to Eastfield for running in, following which the first 24 left for the Continent two months later. One of them, WD No (7)3755, became the 1,000th British-built engine to be transported across the English Channel; to commemorate the event it was named *Longmoor*. This event, however, coincided with Germany's unconditional surrender on 7 May 1945, and No (7)3755 was put into store at Calais together with other 2-10-0s. The remaining engines stayed in Britain, many going to the LNER at March, while others went to the WD at Longmoor.

In Europe early use was made of the engines on Netherlands Railways (NS) from December 1944. In January 1946 60 engines were loaned to NS and numbered 5001 to 5060, although their lack of train-heating made them unpopular in Holland, while their length created problems on the railway's turntables. However, in February, because of a shortage of locos on the German Deutsche Reichsbahn (DR) in the Ruhr, part of the British Zone, 50 of the 2-10-0s were transferred there, including seven of the NS engines. These were released in June and all engines were

Left: In the 1960s LMR engines that required heavy maintenance or overhaul were sent to BR's Eastleigh Works for attention. WD 2-10-0 No 601 *Kitchener* (ex-WD No (7)3797) was found inside the works on 31 August 1963. This engine was delivered too late in 1945 to be transported to the Continent so was put into store at Longmoor until October 1946. It was then taken out of store for use on the LMR until it was scrapped in 1967. Between May 1957 and February 1959 it was loaned to BR.
Frank Hornby

Left: WD No 600 *Gordon* is seen climbing away from Forest Bank station on the LMR's last day of operation at the end of October 1969. The first coach appears to be an ex-SECR 'Birdcage' brake. After closure of the LMR in October 1969 the Transport Trust was set up to establish a railway preservation centre there, based at Liss. However, because of the opposition of local residents the scheme failed.
M. E. J. Deane collection, courtesy of Ian Bennett

urchased by NS, giving it a total of 103, numbered NS 5001 to 5103.

Of the 20 engines that were dispatched to the Middle East at the end of 1943, only four were actually used in the region; the rest were put into store at Suez.

Post-war distribution: within Great Britain
When the war ended in May 1945 the 2-10-0s were still being delivered, although 26 of the last 27 were not required by the WD. As a result, 20 engines were loaned to the LNER between June and August 1945 and allocated to March depot, while six were stored at Longmoor. When the WD 2-8-0s returned from Europe they were preferred by the LNER, so late in 1946 the 2-10-0s were returned to the WD. These were also put into store at Longmoor, although in 1946 the WD reinstated three of its own engines.

In 1947 two of these were loaned to BR so that they could take part in BR's 'Locomotive Exchanges', which took place during 1948. While the 2-10-0s were shown to be better than the WD 2-8-0s, neither class was as efficient as a Stanier 8F. Despite this, 25 2-10-0s (20 ex-LNER and five ex-Longmoor engines) were later sold to BR, together with 733 2-8-0s. All were taken into BR stock in December 1948. After overhaul and renumbering at either BR's Ashford, Eastleigh or Brighton Works, the 2-10-0s were all allocated to Scottish sheds. Kingmoor shed at Carlisle (68A) was then within BR's Scottish Region and received several of the engines, although all but one of these later moved to other Scottish sheds. The remaining engine, No 90673, was still at Kingmoor shed when it became 12A within BR's London Midland Region, the only representative of its class based in England. The last two BR engines numerically both initially carried the name *North British*. The first of the class was withdrawn in 1961, the last in December 1962.

The LMR's long-term resident WD No (7)3651

WD/MOD No 600 *Gordon* in preservation

No 600 *Gordon* is seen shortly after its arrival at the fledgling Severn Valley Railway (SVR) approaching Eardington Halt with a train to Bridgnorth from Hampton Loade, then the terminus of the line. Eardington, the only intermediate station, first opened in February 1862 and was closed by BR in September 1963. It was reopened by the SVR in 1970, but closed again in 1983.
M. E. J. Deane collection, courtesy of Ian Bennett

Since 1971 *Gordon* has spent very little time away from the SVR. It did, however, go north to take part in the 'Grand Parades' for both the 1975 'Shildon 150' and 1980 'Rainhill 150' events. It is seen here at Rainhill in May 1980.

(NB No 25437, 1943) was joined by a second engine, No (7)3797 (NB No 25645, 1943), which was the last of its class still in store there. The engines were numbered 600 and 601 and named *Gordon* and *Sapper* (later *Kitchener*) respectively. The latter engine spent time on loan to BR between May 1957 and February 1959, and was scrapped in 1967, but No 600 remained in use until the LMR closed in October 1969.

Post-war distribution: outside Great Britain

The 2-10-0s purchased by the Dutch NS lasted only until April 1952. Increasing electrification in the Netherlands led to their early withdrawal by 1951 together with the 2-8-0s. The tenders were retained for use with other engines. Of the engines stored in the Middle East, four were hired, then sold to the Syrian Railways (DHP Nos 150.685 to 150.688). In addition, 16 were sent to Greece in January 1946 (HSR Class Lb, Nos 951 to 966). Both countries carried out a number of small improvements to their engines, after which they were used on express passenger duties. In Syria they hauled the 'Taurus Express', which ran from Baghdad to Istanbul between Aleppo and the eastern Turkish border, while in Greece their express work included trains between Alexandroupolis and Pythion on the western Turkish border. Following the introduction of diesels from 1967, all the Syrian engines had been withdrawn by 1976, while in Greece they were largely restricted to freight work. The first was withdrawn in June 1973, the last in October 1976. A number of engines were then put into store.

Preservation

A number of the class have been preserved, as listed in Table 4. These include the LMR's No 600 *Gordon*, which is currently on display at the Engine House

Left: For many years *Gordon* was stored in the open at Arley station on the SVR. When the Engine House museum opened at Highley in March 2008 the engine was cosmetically restored and has been on display there ever since. It is pictured in the museum on 30 April 2019.

'Austerity' 2-10-0s working for BR

Below left: WD No 90766 (ex-WD No (7)3790 (NB No 25636, 1945)) was another of the 20 engines loaned to the LNER in June/July 1945. It returned to the WD in November 1946, was put into store at the LMR, then sold BR. It was overhauled at Brighton Works during 1949/50 and received its BR number in June 1949. It was among the engines allocated to Motherwell shed (66B) before moving to Grangemouth (65F). The engine it seen passing Grangemouth 'light engine' on 14 June 1962. The engine was among the last of its class to be withdrawn, in December 1962.
Frank Hornby

Left: WD No 90773 *North British* (ex-WD No (7) 3798 (NB No 25644, 1945)) approaches South Leith with a heavy coal train on 16 June 1962. The engine was initially put into store at Longmoor;, it then went to WD Cairnryan before being loaned to the LMS and subsequently purchased by BR. It was among the last batch to be withdrawn, in December 1962. On the right is English Electric Type 3 Bo-Bo diesel-electric No D8076 (later Class 20 No 20076). *Frank Hornby*

museum at Highley on the Severn Valley Railway (SVR). Soon after its arrival at the SVR it travelled north to appear at the celebrations to mark the 125th anniversary of the opening of the Stockton & Darlington Railway at Shildon in 1975, followed in 1980 by those to celebrate 125 years of the Liverpool & Manchester Railway at Rainhill near Liverpool.

For a time it appeared that *Gordon* would be the sole survivor of its class in Britain. However, as listed in Table 4, in August 1984 two more ex-WD 'Austerity' 2-10-0s were imported from store in Greece.

Two other Greek examples, Lb Nos 962 and 964 (WD No (7)3677 (NB No 25463, 1943) and No (7)3682 (NB No 25468 of 1944)), were taken out of store at Thessalonica and restored to operate tourist trains. Because of the economic situation in Greece, it appears that they have not been used for some time, although No 962 was reported as being intact in 2019. At least two others are reported to be dumped at Thessalonica or elsewhere in Greece (see Table 4).

Also preserved is NS No 5085 *Longmoor* (WD No (7)3755, NB No 25601, 1945), the 1,000th engine to be ferried to the Continent, which is now at the Dutch National Railway Museum, Utrecht.

Left: BR No 90756 (ex-WD No (7) 3780 (NB No 25658, 1945) was another of the 20 engines loaned to the LNER in June/July 1945. It returned to the WD in September 1946, was put into store at the LMR and then sold to BR. It was overhauled at Ashford works during 1949/50. It was allocated to Motherwell (then BR(ScR) 66B) in 1950 and was withdrawn in December 1962. It is seen, along with another 2-10-0, still at its home depot at Motherwell on 14 April 1963. *John Price*

Above: WD No 90763 (ex-WD No (7)3787 (NB No 25633, 1945)) was another of the 20 engines loaned to the LNER in June/July 1945. It returned to the WD in September 1946 and was put into store at the LMR, then sold to BR. It was overhauled at Eastleigh Works during 1949/50, and allocated to Carlisle Kingmoor (68A) (then Scottish Region) in April 1950, having received its BR number in March. Kingmoor initially had six of these engines, but when Carlisle was transferred to the London Midland Region in May 1958, Kingmoor becoming 12A, it was the only one to remain. It thus became the only example allocated to an English BR region. The engine is pictured here away from its home territory at the joint ex-GWR/LMS engine shed at Birkenhead (6C), while on loan there in 30 August 1959. It was also withdrawn in December 1962. Beyond it is ex-GWR 'Mogul' No 6374. *Frank Hornby*

'Austerity' 2-10-0s in the Netherlands and Greece

Above and above right: The first 2-10-0 to be preserved was WD No (7)3755 *Longmoor* (NB No 25601 of 1945), which was the 1,000th engine to be ferried to the Continent and worked in both Belgium and the Netherlands, remaining there as NS No 5085. It was withdrawn in the early 1950s and set aside for preservation. It is now at the Dutch National Railway Museum, Utrecht. The second picture shows the engine's spacious cab and controls.
Courtesy of the Dutch National Railway Museum, Utrecht

Right: WD 2-10-0 No (7)3653 (NB No 25439, 1943) was one of 20 engines sent to the Middle East in March 1944, but, after being run in, the engines were put into store at Suez. Four were sent to Syria in July 1944 but the others remained in store until sent to Thessalonica in January 1946. There they became Greek Railways (HSR) Class Lb Nos 951 to 966, ex-WD No (7)3653 becoming Lb No 952. The engine was withdrawn in January 1974 and is pictured stored at Thessalonica shed in August of that year. Next to it is Class S160 2-8-0, HSR Class THg No 587 (USATC No 3428 (Baldwin No 70385). The engine came to Greece from Italy where it ran as FS. 736.166 (see page 120). *Colour Rail.*

Right: WD 2-10-0 No (7)3652 (NB No 25438, 1943) was also sent to Greece from store at Suez, becoming HSR Lb No 951. It, too, was withdrawn in January 1974 and, after spending time in store, it was purchased for preservation and went to the Mid-Hants Railway at Ropley, arriving at Ipswich Docks in August 1984. It is seen here in steam at Ropley in LMR Prussian blue livery, which it never carried, masquerading as WD No 601 *Sturdee*, on 15 June 1989. Behind it is ex-USATC Class S160 No 3278 (HSB No THg 575), which had also come from Greece (see Chapter 4).
Ray Ruffell, Slip Coach Archive

Below right: Ex-WD 'Austerity' No (7)3652, restored as ex-LMR WD No 601 *Sturdee*, is seen again approaching Alresford with a train from Alton during the Mid-Hants Gala in July 1989.

Below: No (7)3652 is now at the North Norfolk Railway at Weybourne, where it is running in the guise of BR Class WD 2-10-0 No 90775. The last BR loco in this series was No 90774, which was withdrawn in December 1962. The engine is seen on a visit to the Great Central Railway in June 2002.

Above: Back on the North Norfolk Railway, No 90775 is seen arriving at Weybourne with a train from Sheringham to Holt on 8 May 2018. The engine is now named *Royal Norfolk Regiment*.

Right: The second WD 'Austerity' 2-10-0 to be imported from Greece (HSR No Lb 960, ex-WD No (7)3672 (NB 25458, 1944)), also arrived at Ipswich Docks in August 1984. It went initially to the Lavender Line at Isfield, Sussex, where it was named *Dame Vera Lynn* by the lady herself in August 1985. It then went to the Mid-Hants Railway for overhaul and is seen at Ropley while under test on the 19 August 1986. It then went to the North Yorkshire Moors Railway at Grosmont, where it has been ever since. *Frank Hornby*

Left: WD 'Austerity' 2-10-0 No (7)3672 *Dame Vera Lynn* is seen in action in August 1989 after its arrival at the NYMR. Here the engine and its train have left Goathland and are climbing towards the summit of the line at Fen Bog.

Right: Dame Vera Lynn is seen at Grosmont while running round its train after arrival from Pickering in August 1989. The engine is currently undergoing an extended restoration at Grosmont. *Steve Frost*

3.5. Bulleid's Southern Railway 'Austerity' 0-6-0s
Construction and distribution

O.V. Bulleid became the Chief Mechanical Engineer of the Southern Railway (SR) in 1937. During his tenure, which ended in 1949, he produced four classes of innovative and controversial steam engines. Three of these, the 'Merchant Navy' and 'West Country'/'Battle of Britain' 'Pacifics' together with his ill-fated 'Leader' Class engines, could by no means be described as 'Austerities', even though the 'Pacifics' were first built during the war. In contrast, his Class Q1 0-6-0 was designed and built with the war very much in mind. Full details were given in *The Railway Magazine* of May to June 1942, as seen in the accompanying illustrations.

Although generally considered to be the most ugly engines ever produced in Britain, they were nevertheless the most powerful 0-6-0 ever built in the country and their introduction eliminated much of the double-heading previously needed to cope with the SR's heavy wartime traffic. In all, 40 were built, 20 each at Brighton and Ashford. Bulleid also introduced a new numbering system for his steam engines based on Continental practice, and the Class Q1s were thus numbered C1 to C40, where the 'C' indicated an 0-6-0.

Although designed to operate over the whole of the SR network, they were initially concentrated at

An extract from *The Railway Magazine* for May to June 1942, 'Mr. Bulleid's New 0-6-0 Engine' – the Class Q1 'Austerity' for the Southern Railway. *Reproduced with the permission of Mortons Media Group*

Eastleigh and Feltham. In February 1946, Nos C1 to 18 were allocated to Eastleigh while C19 to 40 were at Feltham. On Nationalisation the class was renumbered 33001 to 33040. By September 1953 some had moved to the sheds at Guildford (70C), Stewarts Lane (73A), Hither Green (73C), Tonbridge (74D) and St Leonards (74E) at Hastings. After the Kent Coast electrification schemes of the early 1960s, the engines were again concentrated at Feltham (70B), Tonbridge, Guildford and Eastleigh (71A). The first to be withdrawn was No 33005 in June 1963, with several more going by the end of the year. The last to be withdrawn, in January 1966, were Nos 33006, 33020 and 33027.

Preservation
An example of the class was listed for preservation as part of the National Collection, and the chosen engine was the first of the class, No 33001, built at Brighton in March 1942. During its career the engine was shedded at Eastleigh, Guildford and Feltham. After withdrawal from Guildford shed in May 1964 it spent some time in store before going to the Bluebell Railway at Sheffield Park. There it initially ran carrying its BR number 33001, before reverting to its original SR number C1. Since 2004 it has been displayed at the NRM at York, still carry its SR number C1.

None of the class went to Barry scrapyard and no others were preserved.

Above: The Class Q1 0-6-0s were introduced on the Southern Railway (SR) in 1942 to meet a wartime requirement for powerful freight engines that could work throughout the UK. Designed by O. V. Bulleid, they were stripped of all non-essential parts to create a truly ugly 'Austerity' engine. No 33009 is seen at Guildford on 9 March 1965, where the shed cleaners have cleaned those parts of its number to recreate its old SR number 'C9'. The engine was built at Brighton in July 1942 and withdrawn in September 1965.
Ray Ruffell, Slip Coach Archive

Right: No 33009 is pictured again, this time entering Guildford station with an engineer's train on 27 March 1965; it is double-heading the train with its shedmate No 33018. The latter engine was built at Ashford in April 1942 and was withdrawn in July 1965. Forty members of the class were built, all operated within the SR and later BR(SR).
M. E. J. Deane collection, courtesy of Ian Bennett

Above: Bulleid's Class Q1 heavy freight engines were rarely seen on passenger duties. An exception was the branch line from Gravesend to Allhallows, where they were used on fill-in turns after working freight trains to the Grain oil refinery. No 33036 stands at Stoke Junction Halt with the 2.47pm train from Gravesend to Allhallows on 24 September 1960. The train is comprised of just two pre-Grouping coaches. No 33036 was built at Ashford in December 1942 and withdrawn in June 1964, by which time the Allhallows branch had already closed, on 4 December 1961. *Terry Gough, Slip Coach Archive*

Left: Many of the Q1s were broken up in South Wales, although none went to Dai Woodham's Barry scrapyard, where they might have survived into preservation. In July 1966 No 33020 awaits its fate at Cashmore's yard in Newport. Behind it stands one of the Swindon-built, English Electric-engined, diesel-electric shunters, a Stanier 'Black 5' 4-6-0 and another Q1. No 33020 was built at Ashford in May 1942 and was one of the last three to be withdrawn in January 1966. The engine spent a few days in the former GWR goods yard at Bath, having 'run hot' while being towed to South Wales.

Above: The first engine of the class, BR No 33001, is the only example to be preserved, and is part of the National Collection. The engine was built at Brighton in March 1942 and withdrawn from Guildford shed in May 1964. It is pictured in September 1980 near Freshfield, soon after its return to steam at the Bluebell Railway, with a train from Sheffield Park to Horsted Keynes.

No 33001 is seen in its original Southern Railway black livery at Sheffield Park in February 1994. It carries its SR number C1, where 'C' indicates an 0-6-0 under the Continental system. Since 2004 the engine has been displayed at the National Railway Museum, York.

4 American War Engines

4.1 The USATC Class S100 0-6-0 tank engines

Construction and initial distribution
Once established, the United States Army Transportation Corps (USATC) soon got things moving. The first significant USATC design as far as Europe is concerned was the 'Standard' Class S100 0-6-0 tank, and 382 were built by Davenport (Dav), Porter and the Vulcan Iron Works (VIW) between 1942 and 1944. They were built to the British, and therefore the Continental, loading gauge, minimising both the material and man-hours required. A photograph and brief description of the engines were included in *The Railway Magazine* of January and February 1943 (see the accompanying extract). Their wheelbase was 10 feet, weight 44 tons 13 cwt and tractive effort (TE) of 21,600lb, compared with the 11-foot wheelbase, 48 tons 5 cwt and TE of 23,870 lb for the WD's Hunslet 0-6-0STs. The WD received just over 100 engines from July 1942, most built by Porter. After tests, several improvements to the design were made at the WD facilities at Longmoor and Melbourne (Derbyshire), where they were lettered 'WD'. Others were initially sent by the Ministry of Fuel and Power (MoF&P) for use in various collieries. The rest were put into store.

In all, more than 200 of the engines came to Britain before shipment to France. Many of them were initially lent to the GWR in South Wales, while others were stored there. The Welsh engines were returned to the USATC in August/September 1944 for dispatch to the Continent. These were followed in late 1944/early 1945 by engines sent direct from the USA. Most

An extract from The Railway Magazine *for January and February 1943: 'American-built 0-6-0 tank engines'. Reproduced with the permission of Mortons Media Group*

worked in France and Belgium and, towards the end of hostilities, were stored en masse, many at Louvain in Belgium.

Others went direct to the Middle East from the USA, where they worked in Iraq, Palestine (much of which is now Israel) and Egypt. Most of these were to remain there. In 1943 five engines were sent to Italy from North Africa, four of which were purchased by the Italian FS (Class 831). Further afield four went to Jamaica and worked on the JGR (Class C), while China received around 20 (Class XK2).

Post-war distribution
After the war 42 WD engines were stored at Newbury. Most of these were selected by the

American Locomotives for Great Britain
SEVERAL locomotives built by the H. K. Porter Co. Inc., of Pittsburgh, Pa., U.S.A., have been delivered recently in this country to the order of the Ministry of Supply. They are 0-6-0 tanks with outside cylinders driving the rearmost pair of coupled wheels; steam distribution is effected by Walschaerts valve gearing, and piston valves 8 in. dia. are employed. Simplicity and ruggedness, easy maintenance and repair or replacement are the dominant features. The basic design permits the use of either bituminous coal or oil fuel; conversion from one to the other requires the usual changes at the firebox and front end, and fuel compartments. The boiler contains 150 tubes 2 in. dia. and 10 ft. long. In working order the locomotive weighs 44 tons 13 cwt.; the water capacity of the side tanks is 1,200 U.S. gal. The coal capacity of the bunker is 1 ton, or if oil is used, 300 U.S. gal. is carried in a deck tank. The tractive force exerted by the locomotive is 21,600 lb. The diameter of the wheels is 4 ft. 6 in.; the wheelbase is 10 ft., and the boiler diameter is 4 ft. 4 in. One of the locomotives is illustrated on page 56.

United Nations Relief & Rehabilitation Administration (UNRRA) for distribution in Eastern Europe, while 14 were purchased by Britain's Southern Railway (SR).

Countries to benefit from UNRRA's activities included Austria, Greece and Jugoslavia. The Austrian ÖBB received ten (Class 989), which were withdrawn in the early 1960s. The 20 engines that went to Greece (HSR Class Da) lasted a little longer, surviving into the 1980s. One hundred and six went to the Jugoslavian JDZ (Class 62), some of which came from the store at Newbury. In addition, in 1956 some 80 new Class 62 engines were built by Djuro Djakovic at Slavonski Brod

(DD) in Jugoslavia (now in Croatia). Most of these were for industrial use, although 23 were built for the JDZ. Some of the industrial engines were still in use early in the 21st century.

The Italian FS found itself with four engines (Class 831), which were all withdrawn during 1953. The French SNCF purchased 77 (Class 030TU), which lasted until the end of steam in the early 1970s. In the Netherlands two were used at the Orange-Nassau Mijnen coal mines, numbered ON-26 (ex-USATC No 4389 (Dav No 2533, 1943)) and ON-27 (ex-USATC No 1948 (Dav No 2513, 1943)).

Back in Britain USATC No 4326 (VIW No 4488, 1943) was loaned to the Southern Railway for a trial in Southampton Docks. So successful was it that 13 engines were purchased from the store at Newbury in 1946 (a 14th engine, USATC No 1261, was purchased for spares). The engines were later numbered BR Nos. 30061 to 30074. All were built by the VIW apart from No 30061 (ex-USATC No 1264 (Porter No 7420, 1942)). Five other Porter-built engines had been intially selected but these were changed to VIW-built machines. At Southampton Docks they replaced a number of ex-LSWR Class B4 0-4-0Ts that dated back to 1891. In addition, other engines were used on industrial lines in Britain, including one (ex-USATC No 1938 (Dav No 2501, 1943)) at the Austin Motor Factory at Longbridge, Birmingham. A further engine (USATC No 4382 (Dav No 2531, 1943)) was returned from France to the WD at Longmoor, where it was named *Major General Frank S. Ross*; this became WD No 300 in 1952 and was scrapped in April 1959.

In 1963 most of the 'USA' Class at Southampton Docks were displaced by Class 07 diesel shunters; some were then used as Departmental engines, replacing the few surviving B4s in the process, and a few survived until after the end of steam on BR(SR) in July 1967.

Preservation in Europe

Those engines preserved in Britain and Europe are listed in Table 5. These include four of BR(SR)'s Class 'USA' engines, while at least two Jugoslavian examples have also been preserved (see Table 5a). Two of the original SNCF engines are preserved in France together with a Porter-built example from Jugoslavia. One of the Dutch coal mine engines, No ON-27, has also survived.

At least two examples are preserved in the former Jugoslavia, one in Croatia, the other in Serbia (see Table 5). It is otherwise difficult to determine which other engines survive there. Until recently examples could still be found in industrial use while many others, both industrial and JDZ (JZ after 1956) engines, were dumped across the country. Some of these may subsequently have been preserved, others broken up. Some are pictured here but are not included in Table 5.

Class 100 0-6-0Ts working as Class 'USA' for BR(SR)

The first of the thirteen ex-USATC Class S100 tanks purchased by the Southern Railway in 1946 became BR No 30061 (USATC No 1264 (Porter No 7420, 1942)). The engine was withdrawn in November 1962 following the introduction of diesel shunters (later Class 07) at Southampton Docks. It was then selected for use by the BR(SR) Engineers Dept.,' and numbered DS 233 as pictured here at Eastleigh shed on 2 January 1965. The engine worked at Redbridge Sleeper Works, Southampton until finally withdrawn in March 1967. It was not preserved. *Frank Hornby*

Above: No 30065 (ex-USATC No 1968 (VIW No 4441, 1943)), one of the class withdrawn in October 1962, is seen at Eastleigh shed (71A) on 12 January 1963. It was to become another of the six engines reinstated as Departmental locos, this one becoming No DS237 *Maunsell*. It was allocated to Ashford Works, Kent, and remained there until finally withdrawn in April 1967. It was subsequently preserved on the K&ESR (see page 110).

Above: In all, 13 'USA' tanks survived the arrival of diesel shunters at Southampton Docks. In addition to the six Departmental engines, others found work as shed pilots or on local shunting duties. One of the latter, No 30072 (ex-USATC No 1973 (VIW No 4446, 1943)), is seen shunting at Guildford shed on 12 March 1965, where it had replaced ex-LSWR Class B4 No 30089, built at Nine Elms in 1892. No 30072 was withdrawn in July 1967 and it now preserved on the K&WVR (see page 111).
Ray Ruffell, Slip Coach Archive

Left: Despite the introduction of the diesels, several of the 'USA' Class engines remained active at Southampton Docks. Two of them, Nos 30067 (ex-USATC No 1282 (VIW No 4380, 1943)) and 30071 (ex-USATC No 1966 (VIW No 4439, 1943)), are seen standing at the small coal stage at Southampton Dock depot (71I) on 22 July 1965. Both engines were withdrawn from Eastleigh shed in July 1967, and neither was preserved. Note the two funnels of the Cunard liner *Queen Elizabeth* in the background. *Ray Ruffell, Slip Coach Archive*

Ex-BR(SR) Class 'USA' tanks preserved in England

Above: 'USA' tank No 30064 (ex-USATC No 1959 (VIW No 4432, 1943)) survived early withdrawal and received a new BR(SR) green livery. It was finally withdrawn in July 1967 and purchased for preservation on the Bluebell Railway in Sussex. It is seen here acting as station pilot at Sheffield Park on 19 October 1980.

Above: For a time the Bluebell Railway's 'USA' tank No 30064 ran in its original guise of WD/USATC No 1959. It is pictured at Sheffield Park on 21 October 2004. *Frank Hornby*

Left: After withdrawal as Ashford Works shunter No DS237 *Maunsell* (ex-BR No 30065) in September 1967, the engine was purchased by Woodham's scrapyard at Barry. However, it got no further than Tonbridge when it, and sister engine No DS238 *Wainwright* (ex-No 30070), 'ran hot' and were left abandoned in the yard. Both were then purchased for preservation on the fledgling Kent & East Sussex Railway. No 30065 was the first to be steamed in 1974, and is seen at Tenterden station with a train for Wittersham Road in August 1989.

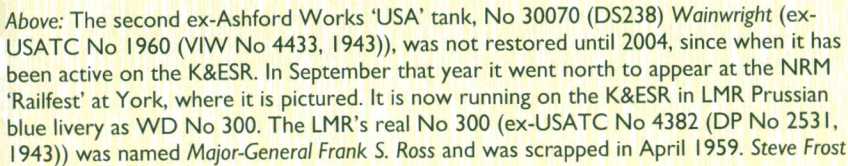

Above: The second ex-Ashford Works 'USA' tank, No 30070 (DS238) *Wainwright* (ex-USATC No 1960 (VIW No 4433, 1943)), was not restored until 2004, since when it has been active on the K&ESR. In September that year it went north to appear at the NRM 'Railfest' at York, where it is pictured. It is now running on the K&ESR in LMR Prussian blue livery as WD No 300. The LMR's real No 300 (ex-USATC No 4382 (DP No 2531, 1943)) was named *Major-General Frank S. Ross* and was scrapped in April 1959. *Steve Frost*

Above right: No 30072 survived as pilot engine at Guildford shed (see page 109) until July 1967. It was then purchased for preservation on the fledgling Keighley & Worth Valley Railway in Yorkshire. The engine was initially given an American appearance with a yellow livery and silver smokebox, and numbered 72. It is seen in this guise outside Haworth's goods shed in August 1969. *Steve Frost*

Right: When pictured in September 1989 at Haworth No 30072 had been repainted in its later BR(SR) green livery, which it carried when pilot engine for Guildford shed in 1966/67.

Ex-USATC S100 tanks around Europe

Above and top right: Two ex-USATC S100 tanks were purchased for work at the Orange-Nassau Mijnen coal mines in the Netherlands They were numbered ON-26 (ex-USATC No 4389 (Dav No 2533, 1943)) and ON-27 (ex-USATC No 1948 (Dav No 2513, 1943)). The former engine was subsequently preserved and is now at the SteamTrain Goes-Borsele, Zeeland. It is pictured at Hoedekenskerke on 7 September 2019 after arrival with a train from Goes.
Courtesy of Reinier Zondervan, Steam Train Goes-Borsele

Right: In 1947 106 ex-USATC Class S100 tanks were selected by UNRRA for use in Jugoslavia, where they became JDZ Class 62. One of them, No 62.096 (ex-USATC No 1258 (Porter No 7414, 1942)), is seen derelict at the extensive loco shed at Nis, Serbia, on 18 September 2001. No 1258 was one of the 42 engines stored at Newbury after 1945 (all but one of the SR's 'USA' tanks came from the same source). The withdrawn electric locos behind the 0-6-0T include Nos 461.008 and 103, built by Craiova during the 1970s.

Above: Another of the ex-JDZ tanks once stored at Newbury was Class 62 No 62.088 (ex-USATC No 1412 (Porter No 7526, 1943)), which survived in industrial use at the Zelvoz wagon works into the 21st century. It is seen here after being coupled to a special train chartered by the LCGB/WRS during their Serbian Tour on 16 September 2001. The special train comprised three four-wheeled coaches and one bogie coach; the latter was detached at the wagon works for renovation. The ramparts of the 15th-century Smederevo Castle can be seen behind the train.

Above right: A further ex-JZD 0-6-0 tank, No 62.046 (ex-USATC No 5041 (Porter No 7584, 1943)), is preserved in France at the Tourist Railway of the Haut Quercy at Martel in the Dordogne, where it is seen on Sunday 19 April 2015. The engine is to be restored as ex-SNCF No 030TU46. It was sent direct from the USA to France late in 1944, the following May it was noted in the dump at Louvain in Belgium before being selected by UNRRA for Jugoslavia. Seventy-seven 'USA' tanks were used by the French SNCF after the war, two of which, Nos 030TU22 (ex-USATC No 4383 (Dav No 2532, 1943)) and 030TU13 (ex-USATC No 6102 (Dav No 7682, 1944)) are preserved in France.

Right: From 1956 some 80 new Class 62 engines were built in Jugoslavia by Djuro Djakovic at Slavonski Brod (DD), and a handful also remained in use into the 21st century. A Serbian factory that continued to use these engines was the Jagodina Wireworks, where their duties included working regular passenger trains for workers at the factory from the JZ station at Jagodina to the Wireworks. Here one of these trains, headed by No 62.377 (DD No 377, 1954), crosses a river between Jagodina and the Wireworks on 19 September 2001. On board were members of the LCGB/WRS's 2001 tour of Serbia.

Above: The LCGB/WRS tour of Serbia also visited the branch line to the Armaments Factory at Dragocevo. Spark-emitting steam engines were banned from the line for obvious reasons, and fireless loco 0-6-0T No LBV No 001 (Sinvoz, 1987) was used instead. The single coach provided for the party had few comforts, such as glass in the windows. The train is seen returning to Dragocevo station on 20 September 2001. The chassis, wheels and cylinders of the engine are from DD-built 'USA' tank No 62.678. At the time of the visit the engine showed signs of shrapnel damage from recent NATO air raids.

Above: The coal mine at Vreoci included both standard and 900mm-gauge railway lines, while some of the latter were electrified. The mine's small fleet of engines included at least two DD-built 'USA' 0-6-0 tanks, which were steamed for the LCGB/WRS tour on 22 September. The two engines, Nos 62.635 (DD No 635, 1955) and 62.127 (DD No 620, 1957) are seen at the head of a rake of coal wagons. To the right is a 900mm-gauge 0-4-0 tank, which was numbered 53.017 but included the frames from No 53.023. The engines were built by the French firm of Decauville in 1953, although it is thought they were actually built under licence. Note also the electrified overhead wire.

Left: Many 'USA' 0-6-0 tanks remained active into the 21st century, including those at the collieries around Tuzla in Bosnia Herzegovina. During an RTC visit to the Bakije Mine depot these recently retired engines were encountered on 14 June 2016. All were stored out of use – and were available for sale! Four of them, headed by No 62.123 (DD No 630, 1957), are seen here, with Nos 62.376, 637 and 368 behind (DD Nos 376, 637 and 368 of 1954, 1955 and 1953 respectively). (Class 62 tanks built by DD for industry kept their works numbers as running numbers.)

Below: An ex-industrial Class 62 is preserved in England. Reliving the days when a Departmental 'USA' Class 0-6-0 tank (No DS 234 (formerly No 30062)) was based at Meldon Quarry in Devon, one of the class arrives at Okehampton station with a Dartmoor Railway train from the Quarry to Sampford Peverell on 28 August 2010. This 'USA' was built new in Jugoslavia for industrial use as No 62.669 in 1960 (DD No 669), and is normally based at the Shillingstone Station Project in Dorset, numbered 30075.

Above: Also stored at the Bakije Mine depot on 14 June 2016 was ex-JDZ 'USA' tank No 62.125 (DD No 614, 1957).

4.2 The USATC Class S160 2-8-0s

Construction and initial distribution

Like the S100 tanks, the USATC Standard Class S160 2-8-0s were built from 1942 to the British, and therefore the Continental, loading gauge, minimising both the material and man hours required. The engines were described in *The Railway Magazine* of March and April 1943, which compared their dimensions to the WD's two classes of 2-8-0. Further details were given in the July and August edition (see the accompanying illustrations). It is thought that around 2,120 were built for use worldwide between 1942 and 1945. These included both coal-and oil-burning versions, together with those of differing gauges for work outside Europe. Three manufacturers were involved: the American Locomotive Co (ALCO), the Baldwin Locomotive Works (BLW) and the Lima Locomotive Works (LIMA).

The first engines arrived in Britain in November 1942. Like many of the engines built for the WD they were not yet needed for the invasion of France, so the first 402 were loaned to the REC for use by the 'Big Four' railway companies. They were distributed as follows: GWR (174), LNER (168), LMS (50) and SR (6). The other four were loaned to the WD, two of which went to the LMR. Around 355 further engines arrived from the summer of 1943 and, like the S100s, were put into store, mainly in South Wales. All of these engines were coal-burners. The stored engines were checked over at railway works around the country, including Newport, Ebbw Junction. The first of these were sent to the Continent in June 1944. By the end of the year the engines on loan in Britain started to leave too. Further engines, including some oil-burners, later went to France and Italy via North Africa or direct from the USA. As civilian order was restored in the previously occupied countries engines were handed over to the French SNCF (560), Belgian SNCB (217) and Italian FS (348). The last five of these were recovered from Rimini Harbour where the boat transporting them had sunk! A sixth engine was also retrieved but used for spares.

Post-war distribution

Both France and Belgium decided against retaining these engines after the war and they were massed into large storage dumps. The future for most of these engines was to lie in Eastern Europe. The engines in Italy became an FS Class 236 Nos. 001 to 348). At the end of 1959 25 of these went to the Greek HSR. Here they joined 27 of their compatriots which were already working there (HSR Class THg Nos 521 to 537 (coal-burners) & Nos 51 to 60 (oil-burners)). The newcomers from Italy, numbered 571 to 595, were coal-burners. In addition 49 of the Class went new to Turkey from the USA in 1943/44 (TCDD Class 45171 Nos. 45171 to 45220). In August 1946 one of the engines, USATC No 3257 (Alco No 71512, 1944), had returned to the LMR where it was numbered WD 700 and named *Major General Carl R. Grey*. It was scrapped in October 1957.

From December 1945 most of the dumped engines were distributed to countries in Eastern Europe under the auspices of the UNRRA. Apart from Austria, these countries were soon to come within the USSR-dominated 'Warsaw Pact' and find themselves behind the Iron Curtain. Approximate numbers are as follows: Austrian ÖBB Class 956 (30), Jugoslavia JDZ Class 37 (65), Czechoslovakia CSD

One of the first of the 2-8-0 American-built "austerity" locomotive
(See page 112)

Above and right: An extract from *The Railway Magazine* for March and April 1943: 'American-built "Austerity" locomotives'. *Reproduced with the permission of Mortons Media Group*

Class 456 (80), and Poland PKP (575). Seventy-five of the Polish engines were provided by UNRRA (PKP Class Tr 201), while 500 appear to have come direct from the USATC (Class Tr 203). In Hungary 510 engines (MAV Class 411) were purchased direct from the USATC (26 of them were kept for spares). In 1955 PKP produced two rather strange rebuilds of its Class Tr 203 engines: two were converted into 4-6-0s (PKP Class Ok 55), and another a 2-8-0 tank (Class TKr 55), but no further engines were rebuilt. The 30 Austrian engines were withdrawn in 1956, but many of the others lasted into the 1970s or '80s. A number were then retained as stationary boilers.

In addition to these countries, the class could also be found working in Algeria, Morocco, Tunisia, India, USSR, North and South Korea, China, Jamaica and the USA, including in Alaska. Five of the Alaskan engines were sold to FC Langreo in Asturias, Spain, in 1958, the only Allied War Engines to work on the Iberian Peninsula.

Preservation in Europe and Turkey

Several of the Class S160s have been preserved, including seven in Britain. These have come from Greece, Poland, Hungary and China (the Chinese example is listed in Table 5a). Britain now has the largest collection of

An extract from *The Railway Magazine* for July and August 1943: 'American-built "Austerity" locomotives'. *Reproduced with the permission of Mortons Media Group*

American-Built "Austerity" Locomotives

CONSIDERABLE interest has been aroused in railway circles by the arrival in this country at the end of last year of the first American-built "austerity" engines for use here and overseas. As our illustration on page 120 shows, these engines have the distinctive appearance of engines in the U.S.A., though from necessity they are on a smaller scale.

The cab is of steel plate construction with the driver's controls arranged for right-hand drive. In the front of the cab on the left side is a door giving access to the high running platforms which are placed along the full length of the boiler on each side of the engine. Steam braking is used on the engine, but both Westinghouse and vacuum automatic brake apparatus are fitted to the engine for train working. Air reservoirs are situated on both sides of the engine under the running platforms.

Buffers and drawgear follow normal British designs. The spring gear, utilising both laminated and coil springs, is compensated through the locomotive. The tender has a capacity of 9 tons of coal (or 1,800 gal. of oil) and 6,500 gal. of water (U.S. measures) and runs of two four-wheel bogies.

Having only two 19 in. cylinders the engine should never be at a loss for steam to fill them. It is understood that the normal operating speed with either goods or troop trains will be of the order of 40 m.p.h.

The annexed table compares the dimensions of the new standard American and British "austerity" 2-8-0s with those of the L.M.S.R. Class "8" 2-8-0s, on which the British design to some extent is based. Certain significant agreements and disagreements will be noted. The American wide firebox, with its grate area of 41 sq. ft., is designed for all descriptions of fuel, and is equal in size to that of a British Pacific, whereas the two British 2-8-0s have 28.6 sq. ft. only. The American engine also has easily the largest superheating surface—480 sq. ft., as compared with 338 sq. ft. in the British "austerity" 2-8-0, and only 215 sq. ft. the L.M.S.R. Class "8" The American 2-8-0 has the largest heating surface, as well as superheating surface and grate area, but the smallest cylinders—19 in. × 26 in. as compared with 19 in. and 19½ in. × 28 in.—and so has the least tractive effort of the three, though all the working pressures (225 lb.) are uniform. Further, the American locomotive is also the shortest of the three, with a wheelbase of 23 ft. 3 in. as compared with the L.M.S.R. locomotive's 26 ft. In appearance the "foreigner" has some strange features, especially the high pitch of the boiler centre line, which leads to a singularly "hunched-up" outline and results in "daylight" appearing between the underside of the boiler and chassis. This enables a wide firebox to be mounted above the rear pairs of coupled wheels. To British eyes the smokebox front is very curious, with its small door hinged on the left, instead of the right, and set eccentrically in the smokebox front to allow room for the air compressor on the right. The long combined sandbox and dome, and the safety-valves set forward of the firebox, are also unusual. It will be interesting to hear what account these "strangers" give of themselves; some of them are already at work here.

LEADING DIMENSIONS OF STANDARD 2-8-0 LOCOMOTIVES

	American "Austerity"	British "Austerity"	L.M.S.R. Class "8"
Cylinders, dia. ...	19 in.	19 in.	18½ in.
Cylinders, stroke ...	26 in.	28 in.	28 in.
Driving wheels, dia. ...	4 ft. 9 in.	4 ft. 8½ in.	4 ft. 8½ in.
Wheelbase, engine ...	23 ft. 3 in.	24 ft. 10 in.	26 ft.
Heating surface ...	1,773 sq. ft.	1,680 sq. ft.	1,650 sq. ft.
Superheating surface ...	480 sq. ft.	338 sq. ft.	215 sq. ft.
Combined heating surface ...	2,253 sq. ft.	2,018 sq. ft.	1,865 sq ft.
Firegrate area... ...	41 sq. ft.	28.6 sq. ft.	28.6 sq. ft.
Working pressure, per sq. in. ...	225 lb.	225 lb.	225 lb.
Tractive effort (at 85 per cent.) ...	31,490 lb.	34,215 lb.	32,440 lb.
Adhesion weight ...	63 tons	62 tons	62 tons
Engine weight (working order) ...	72½ tons	72 tons	70½ tons
Tender, weight capacity ...	6,500 gal.	5,000 gal.	4,000 gal.
Tender, coal capacity ...	9 tons	9 tons	9 tons
Engine and tender weight ...	130½ tons	128 tons	125¼ tons

1 Regulator handle
2 Reversing lever
3 Vacuum automatic brake lever
4 Steam valve, small ejector
5 Steam brake lever
6 Whistle pulls
7 L.H. injector steam valve
8 Blower steam valve
9 Steam sander valve
10 Steam brake valve
11 Steam shut off valve turret
12 Vacuum brake steam valve
13 Mechanical lubricator steam valve
14 Air pump steam valve
15 R.H. injector steam valve
16 Blower control valve
17 Coal watering cock and hose
18 Water gauge
19 Water gauge drain valve
20 Water gauge bottom stop valve
21 Water gauge rod to top stop valve
22 Boiler pressure gauge
23 Westinghouse brake gauge
24 Vacuum brake gauge
25 Firehole door
26 Firegrate rocking levers
27 Injector water and overflow valves
28 Injector operating levers
29 Driver's seat
30 Locker and fireman's seat
31 Flag box
32 Steam sanding lever
33 Three water level test cocks
34 Cylinder cock lever
35 Westinghouse brake lever
36 Steam brake lubricator

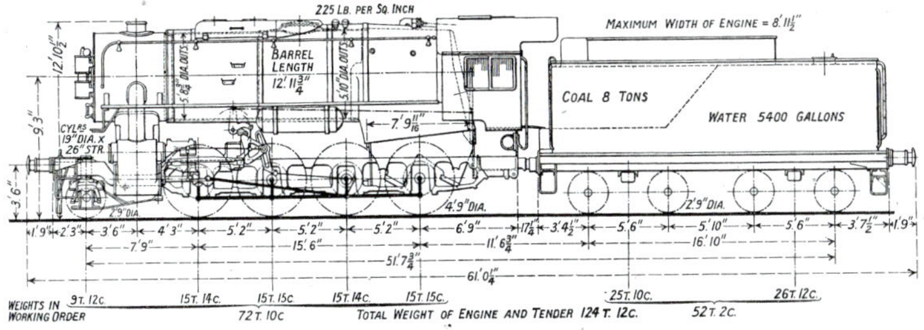

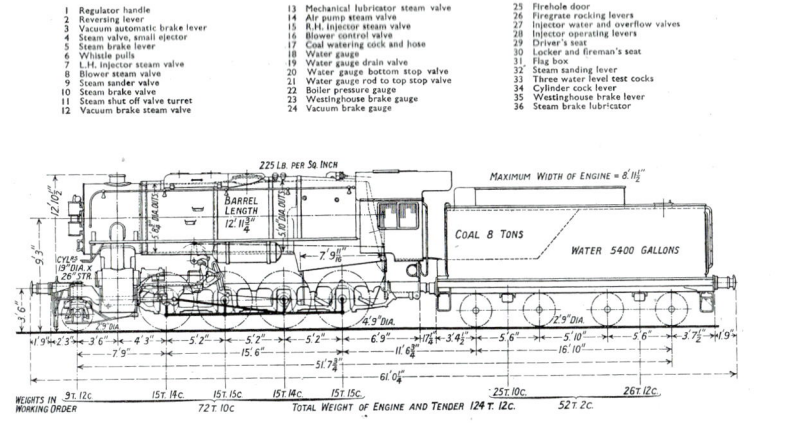

Engine diagram giving principal dimensions and weights

!Cab fittings of U.S.A. "S160" class 2-8-0 locomotive, now at work in this country
(For key to figures, see opposite page)

Some of a group of more than 100 American 2-8-0 austerity locomotives labelled "Transportation Corps, U.S. Army," stored in sidings alongside a double-track main line in England, awaiting their turn for use on the European Continent in connection with the Western Front invasion

the engines outside the USA. As listed in Table 5, others are preserved in Greece and Turkey. Exact numbers are difficult to determine as until recently examples could be found dumped in the various countries. A few in Greece were still intact in 2019 and are included in Table 5.

Above: An extract from *The Railway Magazine* for September and October 1944: American 2-8-0 Austerity locomotives stored somewhere in 'England'. The location is actually in one of the South Wales valleys. *Reproduced with the permission of Mortons Media Group*

Class S160 2-8-0s working for FS in Italy (Class 736) and Turkey (Class 45171)

Right: A total of 244 engines, most of them oil-burners, arrived in Italy in 1944. These later became Italian Railways (FS) Class 736. Here one of them, FS No 736.008 (ex-USATC, No 1732 (Baldwin No 67566)) is pictured at Fermignano on 14 August 1959. This engine was one of those that crossed from North Africa to Italy in the spring of 1944; later engines went direct from the USA. No 736.008 remained in Italy after 1959 when 25 of its fellow engines were sold to Greece. *Ray Ruffell, Slip Coach Archive*

Right: TCDD No 45176 (ex-USATC No 3595 (LIMA No 8508, 1944)) is seen shunting at Kayseri, Turkey, on 2 April 1980. It was one of 50 of its class that went new to the Turkish TCDD, numbered 45171 to 45220.
Brian May

Below: Another Turkish example, TCDD No 45183 (ex-USATC No 1757 ((BLW No 67591, 1942)), hauls a freight out of the yard at Kayseri on the same day.
Brian May

Right: Ex-TCDD No 45215 (ex-USATC No 2210 (LIMA No 8216, 1943)) was one of two of the Turkish TCDD engines to go to the military Kirikkale Munitions Factory as MKE No 45001. The second engine, MKE No 45002, was formerly TCDD No 45210 (ex-USATC No 2118 (ALCO No 70600, 1943)). No 45001 is seen on 8 May 1982 while hauling a workmen's service train using ex-Ottoman Railway carriages. By this time No 45002 had been withdrawn.
Brian May

Class S160 2-8-0s working for HSR in Greece (Class THg) and preserved in the UK

Right: Ex-USATC No 3690 (LIMA 8603, 1944) was one of the engines sent direct from the USA to Italy early in 1945, later becoming FS No 736.199. It was among 25 of its class sent to Greece in late 1959. There they became HSR Class THg, joining 27 that had arrived from the USA in 1947. Several of the class were based at Larissa shed, between Thessalonica and Athens. One of them was No 591, is seen here awaiting its next duty just outside Larissa station on 31 August 1967.

Below: Ex-USATC No 2443 (Alco No 70792, 1943) was one of 27 engines sent direct from the USA to Greece in 1947 where it became HSR Class Thg No 527. It is seen reversing the empty stock of a local passenger train out of the terminus at Thessalonica on 11 August 1972. (See also photos of ex-HSR THg 587 at Thessalonica on page 106 and ex-HSR THg 577, now preserved at Tyseley, on page 121.) *John Price*

Right: In the summer of 1984 three engines arrived at Ipswich Docks from Greece. One was ex-USATC Class S160 No 3278 (ALCO (Canada) No 71533, 1944). This engine had been sent direct from the USA to Italy in the summer of 1944, where it became FS No 736.101. It was sold to Greece at the end of 1959, becoming HSR No THg 577 (its boiler is thought to have come from THg No 575). The other two engines from Greece were both ex-WD 2-10-0s (see pages 101 and 102). USATC No 3278 is seen during restoration at Ropley on the Mid-Hants Railway on 15 June 1989.
Ray Ruffell, Slip Coach Archive

Above: No 3278 is seen again at Ropley on the same day after being named *Franklin D. Roosevelt*. *Ray Ruffell, Slip Coach Archive*

Right: In 1996 Class S160 No 3278 *Franklin D. Roosevelt* moved to the Gloucestershire & Warwickshire Steam Railway. It is seen leaving Toddington with a train to Cheltenham Racecourse in June 1996. After moving to several other locations, No 3278 arrived at the Tyseley Steam Centre, where it is currently being overhauled.

Class S160 2-8-0s in Hungary (Class 411) and preserved in Britain

Right: Hungarian Railways (MAV) 2-8-0 No 411.123 (ex-USATC No 2350 (BLW No 69607, 1943)) shunts in the sidings at Vac on 21 May 1978. After arrival in Britain from the USA in 1943 the engine was loaned to the GWR. It was later returned to the USATC and in October 1944 was shipped to Cherbourg, France, together with hundreds of its compatriots. On the Continent it was one of 217 engines lent to the Belgian SNCB in September 1945. After a period in store it became one of 510 of the class to be purchased by Hungary, becoming MAV Class 411. In Hungary the engines were typically fitted with tall chimneys with spark arrestors and carried a black livery lined out in red. *Brian May*

Left: MAV No 411.264 (ex-USATC No 2781 (LIMA No 8434, 1943)), was discovered by the author at the Kisterenye depot on 17 June 1993. After arriving in Britain in 1943 it was among the many engines stored in South Wales, in this case at Cadoxton. It was overhauled and sent to France in August 1944, and later became another of 484 of its class selected to work for MAV. Note the red star, carried on the smokebox doors of all the European Communist countries. The engine was subsequently cosmetically restored and is displayed at the Hatvan Station, Budapest.

Above: One of two Class S160s restored at Cheddleton on the Churnet Valley Railway, ex-USATC No 6046 (BLW No 72080, 1945), is seen during its restoration on 18 July 2010. After arrival in France from the USA in mid-1945 it worked for a time for the Belgian SNCB until put into store at the end of hostilities. It later became one of the engines that went to Hungary, becoming MAV No 411.144. The other Class S160 restored at Cheddleton is USATC No 5197 (LIMA No 8856, 1945), which was one of 40 engines sent to China in 1947, where it was numbered KD6.463.

Above right: Following its restoration No 6046 visited the Nene Valley Railway near Peterborough. Here it is seen crossing the River Nene as it approaches Wansford station with a train from Peterborough on 25 May 2014.

Right: The summer of 2019 saw the engine at work on the Dartmouth Steam Railway. Here it is seen climbing away from Goodrington Sands with a train from Paignton to Kingswear on 10 June 2019.

Ex-PKP Class Tr203 preserved in Poland

Right: Some 575 Class S160s ran on the Polish PKP. Seventy-five were supplied by UNRRA after the war, becoming PKP Class Tr201. In addition 500 more identical engines came direct from the USATC, becoming PKP Class Tr203. One of the class, PKP No Tr203-451 (ex-USATC No 5801 (LIMA No 8739, 1945)), was preserved in Poland by the Krakov Railway Club. This engine had gone direct from the USA to France in October 1944 and later worked for the French SNCF until put into store at the end of hostilities. It then became one of its class sent by the USATC to work in Poland. After spending time at the Chabowka Railway Museum, where it is seen here on 3 August 1993, the engine is now at the Polish National Railway Museum at Warsaw.

Right: Another Class S160 preserved in Poland is No TR203-296 (ex-USATC No 2438 (ALCO 70787, 1943)). When seen at the Jaworzyna Railway Museum on 1 August 1993 the engine was standing next to one of the last surviving examples of a British-built 'Liberation' Class 2-8-0, No Tr202-28 (ex-UNRRA 1D-92 (VF No 5448, 1946)). The S160 arrived in Britain late in 1942 and early in 1943 was one of 174 engines working for the GWR. After recall by the USATC it went to France in October 1944, where it worked for a time for SNCB at Antwerp. After a period in store it became another of its class sent to Poland from the USATC.

Ex-PKP Class Tr203s preserved in Britain

Right: USATC No 5820 (LIMA No 8758, 1945) arrived in France direct from the USA in mid-1945. There it became an SNCF engine and, after a period in store, was sent by the USATC to Poland, becoming PKP No Tr203-474. It was the first Class S160 to be brought back to GB in the mid-1970s, and was restored at the K&WVR. It is seen here at Oxenhope in May 1986 while running round a train from Keighley.

Below: After an extensive overhaul at the K&WVR, No 5820 makes a dramatic departure from Keighley for Oxenhope on 11 October 2014. *Steve Frost*

Below left: A further PKP engine, No Tr203-288 (ex-USATC No 2253 (BLW No 69496, 1944)), was purchased privately for preservation on the NYMR. The engine originally arrived in Britain late in 1942 and was one of 168 of its class to work for the LNER during 1943. It was recalled by the USATC and sent to France in September 1944. It later worked for the Belgium SNCB at Aarschot and, after a period in store, become yet another of its class to go from the USATC to Poland. After working for a time on the NYMR it was set aside and stored at Grosmont, where it is seen on 2 October 2010. *Steve Frost*

Right: Ex-USATC No 2253 is seen on display at the NRM Shildon on 19 September 2015. Some sources claim the engine is in fact ex-USATC No 2089 (ALCO No 70571, 1943)). It also arrived in Britain in 1943 and was allocated to the LMS at Toton, Nottinghamshire, before transport to France and eventually Poland. It is possible that it contains parts from both engines, although its builder's plate clearly identifies it as USATC No 2253, Baldwin No 69496. *Steve Frost*

No Tr203-288 No 2253 has recently been restored at Grosmont on the NYMR in an attractive maroon livery. Now named *Omaha*, one of the D-Day landing beaches used by the US Army in June 1944, it was pictured outside the shed at Grosmont on 19 July 2019. *David Bate*

Appendix 1: Engine numbering on Europe's railways, 1939-90

The logic to the numbering of steam engines across Britain and Europe during this period was surprisingly diverse, and some of the systems used are outlined below. Britain's 'Big Four' railways companies also used differing systems, but it is assumed that readers will already be familiar with these; the engines were, to an extent, rationalised under BR.

Numbering on German railways

The numbers carried on DRG engines comprised a class number and a running number, e.g. 50.001. In general the lower class numbers were given to express tender engines, followed by mixed-traffic designs. Heavy freight were classed in the 40s and 50s, although these included both old 0-6-0 designs and modern 2-10-0s. Tank engines started in the 60s for express classes and ended with heavy shunting engines in the 90s. Again, many old 0-4-0 and 0-6-0 engines were included. All narrow-gauge steam engines were classed 99.

The German War Engines retained this system with their running numbers being added to those built previously. Thus the Class 50 2-10-0 ran from 50.001 to 50.3171, including those built, from No 50.1583, as ÜK locos. The KDL 'Kriegslok' 2-10-0s were Class 52. Both DB and DR retained these numbers after the war, but during the 1950s DR rebuilt most of its Class 50 and 52 engines to create sub-classes 50.35 and 52.80. Differing computerised numbering was introduced, by DB in January 1968 and DR in July 1970. The DR system mainly involved adding a single computer check number to the end of the running number (e.g. No 50.3014 became 50.3014-3). The DB system in contrast involved greater changes (e.g. No 50. 2146 became 052.146-4). The latter number might be confused with the 'Kriegslok' Class 52s, but by this time DB's engines had long been withdrawn. Most DR stock, including its remaining steam engines, was renumbered in the DB system on 1 January 1992, while Germany was officially reunited on 1 January 1994. The railways were then combined as Deutsche Bahn). Where German engines were subsequently renumbered the original DRG number is given in the text..

WD numbering

The War Department renumbered its steam engines twice, in 1944 and 1952. In 1944 70,000 was added to the WD's three-or four-figure numbers specifically to avoid confusion with numbers in other countries, including Britain (in this book where the '7' was carried it is shown in brackets). By 1952 the WD had disposed of most of its steam engines, so the system was simplified to a series of three digits between 010 and 999. Of the classes described in this book, the 'Austerity' saddle tanks were Nos 100 to 299, the USATC 0-6-0Ts Nos 300 to 399, the WD 2-8-0s Nos 400 to 499, the Stanier 8F 2-8-0s Nos 500 to 599, the WD 2-10-0s Nos 600 to 699, and the USATC 2-8-0s Nos 700 to 799. Apart from the saddle tanks, only the first few numbers were ever required. Higher numbers were given to the WD's diesels. In 1968 only a few steam engines remained in service, most of them saddle tanks. What was by then the MOD therefore reduced the numbers carried by 100, with No 190 becoming No 90, etc.

National Coal Board (NCB)

Many 'Austerity' saddle tanks were later sold to Britain's NCB. Numbering here appears to have been fairly random and in some cases names were used instead; where known these are quoted in the text. In the mid-1970s remaining engines were numbered in the 63000 series. These numbers are not used here.

USATC numbering

The numbers of the USATC's Class S.100 0-6-0 tanks, including those destined for Europe, were numbered as follows: 1252 to 1316, 1387 to 1436, 1927 to 2001, 4313 to 4341, 4372 to 4401, 5000 to 5060, 6000 to 6023, 6080 to 6103 and 6160 to 6183. Numbering of the Class S.160 2-8-0s started at 1600 and, with many gaps, ended with 6078 in 1945.

Other European railways

As in Great Britain, steam engines in **Luxembourg**, the **Netherlands** and **Norway** did not carry their class numbers. Several European railways, like the German DR, started with a class number, followed by the running number; this included **Austria**, which, after its annexation by Nazi Germany in March 1938, renumbered its steam engines within the DR sequence. These numbers were retained by the ÖBB after the war, except for its narrow-gauge engines. The Classes 50 and 52 therefore remained the same, although the ÖBB renumbered its Class 52s built with bar frames as Class 152.

Those countries that used the same system, although not the same number, include the following (the class number of the DR Class 52 is given in each case): **Belgium** Class 26, **Bulgaria** Class 15 and **Jugoslavia** Class 33. Other countries using this system included **Italy** and **Greece**. On the **Italian** FS the ex-WD Stanier 8Fs were FS Class 737 and the ex-USATC Class S160s Class 735. In **Greece** the ex-WD 2-10-0s were Class Lb while the Class S160s were Class Thg.

In other countries the numbers included technical information about the class. Those Class 52s requisitioned by the **USSR** retained their DR running numbers but became Class TE, where the 'T' indicates a war trophy engine (*Trofeinyi*) and the 'E' its power classification. The numbers of the **French** SNCF engines included their wheel arrangement. Those Class 52s that briefly worked in France after the war were numbered 150Y (where 150 indicates a 2-10-0 of Class Y); this is then followed by the running number. A similar system was used in **Romania** for CFR engines built after the First World War. Here the Class 52s were Class 150.1 (some German-built Class 50s were also included in this class), to distinguish them from the Romanian-built Class 50s, which were Class 150. Romanian steam engines also carried other information on metal plates on their cabsides (see the photo on page 45).

On the **Czechoslovakian** CSD the Class 52 became Class 555. The first '5' indicates the number of coupled axles, and the second '5' its maximum speed – rather bizarrely you have to add 3 and multiple by 10 to give 80km/h. The third figure provides its axle loading – if you add 10 – to make 15 tonnes.

The **Hungarian** MAV numbering also involved three numbers. The first digit indicates the number of coupled axles, while the next two show the engine's maximum axle loading. The MAV Class 52s are Class 520. The **Polish** PKP also used a fairly complex system; if the first capital letter of its 'number' is a 'P', it is an express engine (*Pospiean*), while an 'O' is an ordinary passenger engine (*Osobowy*) and a 'T' a freight engine (*Towarowy*). If the second capital letter is a 'K', then it's a tank engine (*Kusy*). These capital letters are followed by a lower-case letter that gives the wheel arrangement: an 'a' is an 0-4-0, a 'z' a 2-10-2. The next two numbers give the year in which the class was first built, if built in Poland. Numbers 1 to 10 and above 100 are engines built abroad. The running number follows. Those Class 52s built outside Poland were therefore Class Ty2, while those built in Poland (despite the country being under German occupation) were Class Ty42.

Appendix 2: Abbreviations used in the text

State railways 1938 to 1994

BBÖ	Bundesbahn Österreich (Austrian State Railways 1923 to 1938, when it became part of the DR)
BDZ	Bulgarski Drzavni Zeleznizi (Bulgarian State Railways)
BR	British Railways
BR(ER)	British Railways (Eastern Region)
BR(LMR)	British Railways (London Midland Region)
BR(SR)	British Railways (Southern Region)
BR(ScR)	British Railways (Scottish Region)
BR(WR)	British Railways (Western Region)
BCh	Belaruskaja Cyhunka (Belarus Railways from 1990s)
CD	Ceske Drahy (Czech State Railway from 1992)
CFL	Société Nationale des Chemins de Fer Luxembourgeois (Luxembourg State Railways)
CFR	Caile Ferate Romane (Romanian State Railways)
CMD	Bohemian-Moravian-Protectorate (when under German control)
CSD	Ceskoslovenske Statni Drahy (Czechoslovakian State Railways to 1992)
DHP	Chemin de fer Damas-Hama et Prolongements (Syrian Railways to 1956)
DRG	Deutsche Reichsbahn Gesellschaft (German State Railway Company to 1949)
DB	Deutsche Bundesbahn (German Federal Railways 1949-93)
DR	Deutsche Reichsbahn (German State Railways 1949-93)
FS	Ferrovie dello Stato (Italian State Railway)

GKB	Graz-Köflacher Bahn, Austria
GySEV	Gyor-Sopron-Ebenfurti Vasut (Austria/Hungary)
HZ	Hrvatske Zeljeznice (Croatian Railways from 1990s)
HDZ	Hrvatske Drzavne Zeljeznice (Croatian State Railways, 1942-45)
HSR	Hellenic State Railway (Greece; OSE from 1971)
IrSR	Iranian State Railway
JDZ	Jugoslovenska Drzavne Zeleznice (Jugoslav State Railways, 1931-56)
JZ	Jugoslovenske Zeljeznice (Jugoslav Railways, 1956-90)
MAV	Magyar Allamvasutak (Hungarian State Railways)
NS	Nederlandse Spoorwegen (Dutch State Railways)
NSB	Norges Statsbaner (Norwegian State Railways)
OSE	Hellenic Railways Organisation (from 1971)
ÖBB	Österreichische Bundesbahnen (Austrian Federal Railways, from 1947)
PKP	Polskie Koleje Państwowe (Polish State Railways)
RZhD	Rossiyskie Zhelezyne Dorogi (Russian State Railways from 2003)
SDZ	Srpske Drzavne Zeleznice (Serbian State Railways, 1942-45)
SKGLB	Salzkammergut-Lokalbahn (Salzkammergut Regional Railway)
SNCB	Société Nationale des Chemins de Fer Belges (Belgian State Railways)
SNCF	Société Nationale des Chemins de Fer Français (French State Railways)
SNCFT	Société Nationale des Chemins de Fer Tunisiens (Tunisian State Railways)
SJ	Statens Jarnvagar (Swedish State Railways)
SZ	Zeleznice Slovenskej (Slovakian Railways (1939-45, and from 1990s)
SZD	Sovetskaja Zeleznaja (Soviet Railways to 1990s)
TCDD	Turkiye Cumhuriyeti Devlet Demiryollari (Turkish State Railways)
UZ	Ukrzaliznytsia (Ukrainian Railways from 1990s)
ZSR	Zeleznice Slovenskej Republiky (Slovakian Railway from 1993)
ZS	Zeleznice Srbije (Serbian Railways from 1990s)

Other railway operators/providers during and after the war

GWR	Great Western Railway (Britain)
LNER	London & North Eastern Railway
LMR	Longmoor Military Railway (Britain)
LMS	London, Midland & Scottish Railway
MEF	Middle East Force (Britain)
MOD	Ministry of Defence (Britain)
MoF&P	Ministry of Fuel & Power (Britain)
MoS	Ministry of Supply (Britain)
MPS	USSR Ministrvo Putej Schrobchenia (USSR Ministry of Transportation)
REC	Railway Executive Committee (Britain)
SR	Southern Railway (Britain)
UNRRA	United Nations Relief and Rehabilitation Administration
USATC	United States Army Transportation Corps
WD	War Department (Britain)

Locomotive builders

AB	Andrew Barclay Sons & Co, Kilmarnock
AEG	Allgemeine Elektricitäts-Gesellschaft AG (German General Electricity Company)
AFB	(Société) Anglo-Franco-Belge
ALCO	The American Locomotive Company, Schenectady, New York (also the Montreal Locomotive Works, Canada)
Bat	Société de Construction des Locomotives Batignolles, Paris
BL	Baldwin Locomotive Works, Eddystone, Pennsylvania
BLW	Borsig Lokomotive Werke, Berlin (Borsig Locomotive Works)
BMAG	Berliner Maschinenbau/Schwartzkopff, Berlin
BP	Beyer, Peacock & Co, Gorton, Manchester
CKD	Ceskomoravska-Kolben-Danek, Prague
Dav	Davenport Locomotive Works, Davenport, Iowa
DD	Djuro Djakovic, Slavonski Brod, Jugoslavia (now Croatia)
DWM	Deutsche Waffen und Munitionsfabriken (Cegielski at Poznan, Poland, 1846 to 1939, production for Poland resumed in 1945)
Fives	Compagnie de Fives-Lille, Lille, France
Frichs	Horsens, Denmark
Graff	Elsässische Maschinenbau-Gesellschaft, Graffenstaden/SACM (see below)
Hainaut	Metal works of Hainaut, Belgium (better known as Couillet)
HC	Hudswell, Clarke & Co, Leeds, West Yorkshire
HE	Hunslet Engine Co, Leeds, West Yorkshire
Krenau	Krenau, Oberschlesische Lokomotivwerke (Fablok, Chrzanow, Poland, 1919 to 1939, production for Poland resumed in 1945)
LEW	Lokomotivbau-Elektrotechnische Werke, Berlin (formerly BLW, Berlin)

LIMA	Lima Locomotive Works, Ohio
LKM	Karl Marx Loco Works, Babelsburg, Berlin, GDR (1948-92, formerly MBA)
MBA	Maschinenbau und Bahnbedarf, Babelsburg, Berlin (1941-45, formerly O&K)
MBK	Maschinenbau-Gesellschaft, Karlsruhe
NB	North British Locomotive Co, Glasgow
Ostrowice	Polish town where KDL locos were manufactured during WWII
O&K	Orenstein & Koppel (to 1939 and from 1949 at Dortmund, DBR)
Port	H. K. Porter Locomotive Works, Pittsburgh
RSH	Robert Stephenson & Hawthorns Ltd, Darlington
SACM	Société Alsacienne de Constructions Mécaniques (Alsatian Mechanical Engineering Company, Mulhouse, Alsace, France (see also Graff, above))
SFCM	Société Française de Constructions Mécaniques/Cail, Denain, France
Schneider	Schneider – Creusot at Le Creusot (Saône-et-Loire), France
Skoda	Skoda, Pilzen, Czechoslovakia
VF	Vulcan Foundry Ltd, Newton-le-Willows Britain
VIW	Vulcan Iron Works, Wilkes-Barre, Pennsylvania
WLF	Wiener Lokomotivfabrik Floridsdorf, Vienna

Other abbreviations

APEMVE	L'Association pour la Préservation et l'Entretien du Matériel à Voien Etroite (The Association for the Preservation and Maintenance of Narrow Gauge Railways, Saint-Germain-d'Arcé, France)
CME	Chief Mechanical Engineer
Dzherelo	Ukrainian Railway Tour Company, now defunct
ELNA	Engerer Lokomotiv-Normen-Ausschuss (Locomotive Standards Committee, Germany)
EFZ	Eisenbahnfreunde Zollernbahn (Friends of the Zollern Railway)
FRG	Federal Republic of Germany (West Germany 1949 to present day)
GB	Great Britain
GCR	Great Central Railway
GCR-N	Great central Railway - Nottingham
GDR	German Democratic Republic (East Germany 1949 to reunification in 1990)
IGE	Erlebnisreisen und Reiseservice (Adventure Tours and Travel Service)
KDL	Kriegsdampflokomotiven (War Steam Locomotives)
KHKD	Klub Historie Zeleznicni Dopravy (Railway Transport History Club, Prague)
ÖeGEG	Österreichische Gesellschaft für Eisenbahngeschichte (Austrian Railway History Society)
LCGB	Locomotive Club of Great Britain
L&SWR	London & South Western Railway
NCB	National Coal Board (Britain)
NYMR	North Yorkshire Moors Railway, Grosmont, North Yorkshire
Plandampf	Events organised by DR/DB when steam replaces diesel on scheduled services
PTG	Portuguese Traction Group (tours)
KDL	Kriegsdampflokomotiven (War Steam Locomotives)
K&ESR	Kent & East Sussex Railway, Rolvenden, Kent
K&WVR	Keighley & Worth Valley Railway, Haworth, West Yorkshire
LCGB	Locomotive Club of Great Britain
RomSteam-Aldo	Romanian Railway Tour Company, now defunct
RTC	Railway Touring Club
RTCS	Railway Travel & Correspondence Society
SBS	Schoeller-Bleckmann Steelworks, Vienna
S&DJR	Somerset & Dorset Joint Railway
SVR	Severn Valley Railway
TEFS	To Europe/Everywhere for Steam
Traditionslok	Preserved DR/DB engine kept in working order and used on special trains
USSR	Union of Soviet Socialist Republics
ÜK	Übergangskriegslokomotiven (Transitional War Locomotives) (To avoid confusion Britain) is given instead of United Kingdom
WRS	Warwickshire Railway Society

Appendix 3: Bibliography and other references

1. Books

ABC British Railways Locomotives, Ian Allan, London (1956 & 1962/63 editions)
ABC Locoshed Book, Ian Allan, London (1959 & 1961 editions)
Allied Military Locomotives of the Second World War, R. Tourret, Tourret Publishing, Abingdon (1995 edition)
Atlas Lokomotiv 1, Inz. Jindrich Bek & Karel Kvarda, Nadas, Prague, 1970
Austerity Saddle Tank Locomotives, Brian Reed et al, Industrial Railway Society, 2010
Benelux Railways Locomotives & Coaching Stock, David Haydock, Peter Fox & Brian Garvin, Platform 5, 1994
British Railway Steam Locomotives 1948-1968, Hugh Longworth, Oxford Publishing Co (2005 edition)
Continent, Coalfield and Conservation, A. P. Lambert and J. C. Woods, Industrial Railway Society, 1991
Das Heizhaus Eisenbahnmuseum, Strasshof, Rupert Gansterer, Vienna, 1999
Die Eisenbahnfahrzeuge des Verkehrsmuseums Dresden, Anon, UniMedia, 1997
European Report: Motive Power Listing Hungary, Neil Webster, Metro, 1997
German Railways: Locomotives & Multiple Units, Brian Garvin and Peter Fox, Platform 5, 1993
Hungarian Railways P. M. Kalla-Bishop, David & Charles, 1973
ÖBB/Austrian Federal Railway: Locomotives & Multiple Units, Brian Garvin and Peter Fox, Platform 5, 1989
Observer's Book of Railway Locomotives of Britain, H. C. Casserley, Frederick Warne, London, 1960
Preserved Steam Locomotives of Poland, S. R. Mazurek, P. E. Waters & Associates, 1992
Preserved Steam Locomotives of Western Europe Vols. 1 & 2, P. Ransome-Wallis, Ian Allan, 1971
Preserved Locomotives of British Railways, Peter Hall and Peter Fox, Platform 5, 2007
SNCF/French National Railways: Locomotives & Multiple Units, Brian Garvin, Peter Fox and Chris Appleby, Platform 5, 1986
Soviet Locomotive Types, A. J. Heywood and I. D. C. Button, Frank Stenvalls Forlag, Malmo, 1995
Steam in Europe, P. B. Whitehouse, Ian Allan, 1966
Steam in Turkey, E. Talbot, The Continental Railway Circle, 1981
Taschenbuch Deutsche Dampflokomotiven, Horst J. Obermayer, Franskh'sche Verlag, Stuttgart 1969
The German Class 52 'Kriegslok', Peter Slaughter, Alexander Vassiliev, Roland Beier, Frank Stenvalls Forlag, 1996
The Ironstone Quarries of the Midlands Part VIII, South Lincolnshire, Eric Tonks, Runpast Publishing, Cheltenham, 1991
The Last Steam Locomotives of Western Europe, P. Ransome-Wallis, Ian Allan, 1963
The Locomotives of Jugoslavia, Vol 1, C. J. Halliwell & Associates, Frank Stenvalls Forlag, 1973
The Steam Locomotives of Czechoslovakia, Paul Catchpole, self-published, 1995
The Steam Locomotives of Eastern Europe, A. E. Durrant, David & Charles, 1966
Verzeichnis der Deutschen Lokomotiven 1923-1963, Helmut Griebl & Fr. Schadow, Transpress VEB, 1965

2. Magazines

Extracts from contemporary copies of *The Railway Magazine*, 1939 to 1946 (reproduced with the permission of Mortons Media Group)

3. Websites

Those consulted include the following:

https://www.bundesbahn.net
https://www.dampflok.at
https://www.dampflokomotivarchiv.de
https://www.heeresfeldbahn.de
https://en.wikipedia.org › wiki › List_of_preserved_Hunslet_Austerity_0-6-0ST locomotives
https://www.reichsbahndampflok.de
https://www.steamlocomotive.info
https://www.uklocos.com
https://en.wikipedia.org/wiki/List_of_preserved_steam_locomotives_in_Germany
https://www.railuk.info/steam/steamshed_search.php

Tables

Table 1
Selected ex-DRG Class 86, 44, 50, 52 and 42 engines preserved in Germany

* locos rebuilt by the DR in 1960s as Classes 50.35 or 52.80

DR/DB Nos (pre-computerisation)	DRG No	Other No(s)	Builder	Number	ÜK/KDL	Date	Wheels	Comments
Berlin Technical Museum								
DB 50.001	50.001	-	Henschel	24355	-	1939	2-10-0	First of the class; built before the outbreak of war
DR 52.4966	52.4966	-	MBA	14036	KDL 1	1944	2-10-0	-
Bavarian Railway Museum at Nördlingen								
DB 44.381	44.381	-	Esslingen	4446	-	1941	2-10-0	Oil-burning from 1960
DR 44.546	44.546	-	Krauss-Maffei	16151	-	1941	2-10-0	Oil-burning from 1966 to 1982
DB 50.778	50.778	-	Henschel	25862	-	1941	2-10-0	-
DB 50.995	50.995	-	MBA	13534	-	1941	2-10-0	-
DR 50.3502*	50.481	-	Krauss-Maffei	15832	-	1940	2-10-0	Oil-burning from 1971 as No 50.0072
DR 50.3600*	50.775	-	Henschel	25859	-	1941	2-10-0	-
DR 50.4073	50.4073	-	LKM	124073	-	1960	2-10-0	-
DR. 52.2195	-	-	Henschel	27046	KDL 1	1943	2-10-0	-
-	52. 5348	SZD TE.5348/PKP Ty2-1230	Krenau/Chrzanow	1378	KDL 1	1944	2-10-0	-
-	(42.2768)	BDZ 16.16	WLF	17564	KDL 3	1949	2-10-0	Never carried its DRG No

DR/DB Nos (pre-computerisation)	DRG No	Other No(s)	Builder	Number	ÜK/KDL	Date	Wheels	Comments
Railway Museum at Bochum-Dahlhausen								
DB 44.1377	44.1377	-	Krupp	2799	ÜK	1942	2-10-0	-
DB 50.3075	50.3075	-	MBA	14201	ÜK	1943	2-10-0	-
Railway Museum at Altenbeken								
DB 44.1681	44.1681	-	Schichau	3633	ÜK	1942	2-10-0	Oil-burning from 1960
Dresden Technical Museum								
DR 86.606	86.606	-	BLW	15279	ÜK	1942	2-8-2T	Vogtland Railway Association at Adorf, Saxony
DR 52.9900	52.4900	-	MBA	13970	KDL 1	1943	2-10-0	Converted to burn pulverised brown coal Now at Halle Depot Museum
South German Railway Museum at Heilbronn								
DR 86.457	86.457	-	DWM/ Cegielski	442	ÜK	1942	2-8-2T	Ex-DB Museum loco, damaged in 2005 fire at Nuremberg; now at Heilbronn for cosmetic restoration
DR 44.1315	44.1315	-	Krupp	2737	ÜK	1942	2-10-0	Oil-burning from 1966 to 1982 Now displayed outside the Märklin Museum, Göppingen, Baden-Württemberg
DR 44.1378	44.1378	-	Krupp	2800	ÜK	1942	2-10-0	Oil-burning from 1966 to 1982
DB 44.1489	44.1489	-	Schneider	4731	ÜK	1943	2-10-0	Oil-burning from 1960
DR 44.1616	44.1616	-	Krenau/ Chrzanow	1104	ÜK	1943	2-10-0	Oil-burning from 1966 to 1982
DB 50.3031	50.3031	-	Esslingen	4522	ÜK	1942	2-10-0	-
DR 52.8098*	52.3420	-	Krauss-Maffei	16546	KDL 1	1943	2-10-0	-

DR/DB Nos (pre-computerisation)	DRG No	Other No(s)	Builder	Number	ÜK/KDL	Date	Wheels	Comments
Private railway museum at Hermeskeil								
DR 44.167	44.167	-	BMAG	10983	-	1938	2-10-0	Oil-burning from 1966 to 1982
DR 44.177	44.177	-	Krupp	1997	-	1940	2-10-0	Oil-burning from 1966 to 1982
DR 44.196	44.196	-	Krupp	2018	-	1940	2-10-0	-
DR 44.264	44.264	-	Schichau	3390	-	1939	2-10-0	Built before war; has no boiler
DB 44.434	44.434	-	Henschel	26043	-	1941	2-10-0	Oil-burning from 1966 to 1982
DR 44.500	44.500	-	Krauss-Maffei	16105	-	1941	2-10-0	Oil-burning from 1966 to 1982
DR 44.635	44.635	-	Schichau	3460	-	1941	2-10-0	Oil-burning from 1966 to 1982
DR 44.1040	44.104	-	WLF	9396	ÜK	1942	2-10-0	Oil-burning from 1966 to 1982
DR 44.1056	44.1056	-	WLF	9412	ÜK	1942	2-10-0	Oil-burning from 1966 to 1982
DR 44.1106	44.1106	-	BLW	15155	ÜK	1942	2-10-0	Oil-burning from 1966 to 1982
DR 44.1251	44.1251	-	BLW	15237	ÜK	1942	2-10-0	Oil-burning from 1966 to 1982
DR 44.1412	44.1412	-	Schichau	3604	ÜK	1942	2-10-0	Oil-burning from 1966 to 1982
DR 44.1537	44.1537	-	BLW	15376	ÜK	1942	2-10-0	Oil-burning from 1966 to 1982
DB 50.607	50.607	-	Henschel	25826	-	1940	2-10-0	-
DB 50.1446	50.1446	-	Henschel	26256	-	1941	2-10-0	-
DB 50.1832	50.1832	-	BMAG	11730	-	1941	2-10-0	-
DR 50.3014	50.3014	-	Esslingen	4505	ÜK	1942	2-10-0	-
DR 50.3553*	50.235	-	Krauss-Maffei	15754		1939	2-10-0	Built before war
DR 50.3555*	50.2995	-	WLF	9582	ÜK	1942	2-10-0	-
DR 50.3649*	50.2876	-	BMAG	11932	ÜK	1942	2-10-0	-
DR 50.3662*	50.1249	-	WLF	9183	-	1941	2-10-0	-
DR 52.1423	52.1423	-	Esslingen	4609	KDL 1	1943	2-10-0	-
DR 52.1662	52.1662	-	SACM	7929	KDL 1	1943	2-10-0	-
DR 52.2093	52.2093	-	Henschel	26849	KDL 1	1943	2-10-0	-

DR/DB Nos (pre-computerisation)	DRG No	Other No(s)	Builder	Number	ÜK/KDL	Date	Wheels	Comments
DR 52.6721	52.6721	-	WLF	16172	KDL 1	1943	2-10-0	-
DR 52.8006*	52.2644	-	Henschel	27822	KDL 1	1944	2-10-0	-
DR 52.8090*	52.7778	-	MBA	14362	KDL 1	1944	2-10-0	-
DR 52.8113*	52.1159	-	DWM/ Cegielski	573	KDL 1	1943	2-10-0	-
DR 52.8120*	52.2652	-	Henschel	27830	KDL 1	1944	2-10-0	-
DR 52.8123*	52.1633	-	Graffenstaden	7900	KDL 1	1943	2-10-0	-
DR 52.8197*	52.7581	-	WLF	16929	KDL 1	1944	2-10-0	-
—	(42.2754)	BDZ 16.15	WLF	17640	KDL 3	1949	2-10-0	Never carried its DRG No

Railway Museum at Neuenmarkt-Wirsberg

DB 86.283	86.283	-	MBA	12941	-	1936	2-8-2T	Built before war
DB 44.276	44.276	-	Krauss-Maffei	15745	-	1940	2-10-0	-
DB 50.975	50.975	-	Krupp	2340	-	1941	2-10-0	-
DR 50.3690*	50.1465	-	Henschel	26275	-	1941	2-10-0	-
—	52.5804	ÖBB 52.5804	Schichau	4101	KDL 1	1944	2-10-0	

German National Railway Museum at Nuremberg; some engines currently elsewhere

DR 86.001	86.001	-	MBK	2356	-	1928	2-8-2T	First of its class; built before war; ex-DR Museum loco
DB 44.508	44.508	-	Krauss-Maffei	16113	-	1941	2-10-0	Now on loan to the steam-engine friends at Westerwald
DR 44.1093	44.1093	-	WLF	9449	ÜK	1942	2-10-0	Ex-DR Museum loco; oil-burning from 1966 to 1982 Now on loan to the steam-engine friends at Westerwald
DB 50.622	50.133	SNCB 25.014	BLW	14864	-	1940	2-10-0	Number changed in 1958. Ex-DB Museum loco, badly damaged in fire at the Railway Museum in 2005. Future now uncertain

DR/DB Nos (pre-computerisation)	DRG No	Other No(s)	Builder	Number	ÜK/KDL	Date	Wheels	Comments
DR 50.849	50.849	-	Krauss-Maffei	16058	-	1940	2-10-0	Ex-DR Museum loco. Now part of the DB Traditionslok fleet at Glauchau, Saxony
DR 52.6666	52.6666	-	Skoda	1492	KDL1	1943	2-10-0	Ex-DR Museum loco. Now part of the DB Traditionslok fleet at Schöneweide, Berlin
Railway & Technical Museum at Prora								
DR 44.397	44.397	-	Henschel	26006	-	1941	2-10-0	Oil-burning from 1966 to 1982
DR 50.3703*	50.877	-	Krauss-Maffei	16087	-	1941	2-10-0	-
DR 52.8190*	52.2887	-	Henschel	28244	KDL 1	1945	2-10-0	-
Railway Museum at Stassfurt								
DR 44.663	44.663	-	BLW	15119	-	1941	2-10-0	Oil-burning from 1966 to 1982
DR 44.1182	44.1182	.	Krupp	2684	ÜK	1942	2-10-0	Oil-burning from 1966 to 1982
DR 44.1486	44.1486	.	Schneider	4728	ÜK	1943	2-10-0	Oil-burning from 1966 to 1982
DR 50.3556*	50.1489	-	Henschel	26299	-	1941	2-10-0	-
DR 50.3695*	50.1066	-	BMAG	11555	-	1941	2-10-0	-
DR 52.8137*	52.5803	-	Schichau	4100	KDL 1	1944	2-10-0	-
DR 52.8161*	52.5519	-	Schichau	3796	KDL 1	1943	2-10-0	-
DR 52.8184*	52.1519		WLF	17266	KDL1	1944	2-10-0	
DR 52.8189*	52.5306		Krenau	1327	KDL 1	1944	2-10-0	
Saxon Railway Museum at Chemnitz-Hilbersdorf								
DR 44.1338	44.1338	-	Krupp	2760	ÜK	1941	2-10-0	Oil-burning from 1966 to 19822
DR 50.3628*	50.2678	-	Esslingen	4489	ÜK	1942	2-10-0	-
DR 50.3648*	50.967	-	Krupp	2332	-	1941	2-10-0	-
DR 52.4924	52.4924	-	MBA	13994	KDL 1	1943	2-10-0	-

DR/DB Nos (pre-computerisation)	DRG No	Other No(s)	Builder	Number	ÜK/KDL	Date	Wheels	Comments
DR 52.8068*	52.3311	-	Jung	11322	KDL 1	1944	2-10-0	-
Technical Museum at Speyer								
-	50.685	ÖBB 50.685	WLF	3405	-	1940	2-10-0	Sold to GKN in 1972
-	52.3915	SZD TE.3915	MBA	14169	KDL 1	1944	2-10-0	-
-	42.1504	PKP Ty3-3/ TY43-127	Esslingen	4874	KDL 3	1944	2-10-0	Built in Germany but later fitted with Polish boiler
Zollernbahn Railway Society at Tübingen								
-	52.7596	ÖBB 52.7596	WLF	16944	KDL 1	1944	2-10-0	-

Table 2
Selected ex-DRG Class 86, 44, 50, 52 and 42 engines preserved outside Germany

* locos rebuilt by the DR in the 1960s as Classes 50.35 or 52.80

DR/DB Nos (pre-computerisation)	DRG No	Other Nos	Builder	Number	ÜK/KDL	Date	Wheels	Comments
Austria								
Railway Museum at Strasshof near Vienna								
-	52.100	JDZ 33.044	Krauss-Maffei	16411	KDL 1	1943	2-10-0	-
-	52.460	JDZ 33.240	BLW	15557	KDL 1	1943	2-10-0	-
-	52.7593	ÖBB 52.7593	WLF	16941	KDL 1	1944	2-10-0	-
-	52.7594	ÖBB 52.7594	WLF	16942	KDL 1	1944	2-10-0	-
-	(42.2708)	ÖBB 42.2708	WLF	17591	KDL 3	1946	2-10-0	Built after war; never carried DRG No

DR/DB Nos (pre-computerisation)	DRG No	Other Nos	Builder	Number	ÜK/KDL	Date	Wheels	Comments
Austrian Railway History Society (ÖeGEG) at Ampflwang								
DR 86.056	86.056	-	BLW	14428	-	1932	2-8-2T	Built before war
-	86.476	ÖBB 86.476	DWM	461	ÜK	1943	2-8-2T	-
DR 86.501	86.501	-	Henschel	26720	-	1942	2-8-2T	-
-	44.661	-	BLW	15117	-	1941	2-10-0	Oil-burning from 1966 to 1982
-	44.1595	-	SACM	7865	ÜK	1943	2-10-0	Oil-burning from 1966 to 1982
-	44.1614	-	Krenau/Chrzanow	1102	ÜK	1943	2-10-0	Oil-burning from 1966 to 1982
-	50.499	-	BLW	14947	-	1940	2-10-0	-
DR 50.3506*	50.903	-	Krupp	2364	-	1940	2-10-0	-
DR 50.3519*	50.342	-	Henschel	24976	-	1940	2-10-0	-
DR 50.3689*	50.547	-	Henschel	25766	-	1940	2-10-0	-
-	52.1198	ÖBB 52.1198	DWM	612	KDL 1	1943	2-10-0	-
-	52.3316	ÖBB 52.3316	Jung	11327	KDL 1	1944	2-10-0	-
-	52.3517	ÖBB 52.3517	Krauss-Maffei	16643	KDL 1	1943	2-10-0	-
-	52.4552	ÖBB 52.4552	DWM	868	KDL 1	1944	2-10-0	-
-	52.7102	ÖBB 52.7102	WLF	16555	KDL 1	1944	2-10-0	-
DR 52.8003*	52.6357	-	BMAG	12810	KDL 1	1944	2-10-0	-
DR 52.8096*	52.2312	-	Henschel	27480	KDL 1	1943	2-10-0	-
DR 52.8124*	52.2501	-	Henschel	27669	KDL 1	1943	2-10-0	-
DR 52.8186*	52.632	-	Schichau	4110	KDL 1	1944	2-10-0	-
DR 52.8196*	52.5374	-	Krenau/Chrzanow	1407	KDL 1	1944	2-10-0	-
-	(42.2750)	BDZ 16.19	WLF	17636	KDL 3	1949	2-10-0	Never carried its DRG No

DR/DB Nos (pre-computerisation)	DRG No	Other Nos	Builder	Number	ÜK/KDL	Date	Wheels	Comments
Brenner & Brenner Steam Locomotive Operating Co, Vienna								
-	50.1171	ÖBB 50.1171	Skoda	1250	-	1941	2-10-0	Sold to GKN in 1972
-	52.1227	ÖBB 52.1227	DWM	654	KDL 1	1944	2-10-0	-
-	52.7612	ÖBB 52.7612	WLF	16960	KDL 1	1944	2-10-0	-

Belgium

Operated by Heritage Rail Tours of Brussels

DR/DB Nos	DRG No	Other Nos	Builder	Number	ÜK/KDL	Date	Wheels	Comments
-	52.3554	SZD TE.3554 PKP Ty2.3554	Krauss-Maffei	16691	KDL 1	1943	2-10-0	Now running as SNCB 26.101

Steam Railway of the Three Valleys at Mariembourg

DR/DB Nos	DRG No	Other Nos	Builder	Number	ÜK/KDL	Date	Wheels	Comments
DR 50.3696*	50.193	-	Krupp	2059	-	1939	2-10-0	Built before war
	52.3316	ÖBB 52.3316	Jung	11325	KDL 1	1944	2-10-0	-
DR 52.8200*	52.467	-	BLW	15564	KDL 1	1943	2-10-0	-

Preserved at Schaerbeek

DR/DB Nos	DRG No	Other Nos	Builder	Number	ÜK/KDL	Date	Wheels	Comments
-	52.7173	SZD TE.7173 PKP Ty2-7173	WLF	16626	KDL 1	1943	2-10-0	Formerly at Nene Valley Railway, Wansford GB

DR/DB Nos (pre-computerisation)	DRG No	Other Nos	Builder	Number	ÜK/KDL	Date	Wheels	Comments
Bulgaria								
Asenovo Depot								
–	52.2728	BDZ 15.17	Henschel	27969	KDL 1	1944	2-10-0	Strategic Reserve
–	52.3240	BDZ 15.24	Jung	11251	KDL 1	1944	2-10-0	Stategic Reserve
Plovdiv depot								
-	50.463	BDZ 14.25	WLF	3328	-	1940	2-10-0	-
Gorna Oryahovitse Depot								
-	52.5658	BDZ 15.131	Schichau	3936	KDL1	1943	2-10-0	Strategic Reserve, seen 2019
-	(42.2761)	BDZ 16.01	WLF	17647	KDL 3	1949	2-10-0	Never carried its DRG No
-	(42.2746)	BDZ 16.27	WLF	17632	KDL 3	1948	2-10-0	Never carried its DRG No
Countries of the former Czechoslovakia								
Czech Railway Museum at Luzna								
-	52.7620	SZD TE.7620 CSD 555.0153 JDZ 33.502	WLF	1696	KDL 1	1944	2-10-0	Owned by KHKD
Slovak Railway Museum at Bratislava								
-	-	HDZ 30.017 JDZ 33.032	Henschel	27938	KDL 1	1944	2-10-0	Built new for HDZ; formerly at Strasshof, Austria

DR/DB Nos (pre-computerisation)	DRG No	Other Nos	Builder	Number	ÜK/KDL	Date	Wheels	Comments
-	52.7047	SZD TE.7047 CSD 555.0222	WLF	16500	KDL I	1943	2-10-0	3000 added to CSD No when converted to oil-burning in 1960s
-	52.7447	SZD TE.7447 CSD 555.008	Skoda	1523	KDL I	1943	2-10-0	3000 added to CSD No when converted to oil-burning in 1960s

France

The Tourist Railway of the Aa Valley (CFTVA)

-	52.6690	SZD TE.6690 PKP Ty2.6690	WLF	16121	KDL I	1943	2-10-0	-
-	52.332	PKP Ty2.993	Jung	11331	KDL I	1944	2-10-0	-

The Haut Quercy Tourist Railway at Martel

DR 50.3661*	50.1224	-	WLF	9158	-	1941	2-10-0	-

The Pontarlier-Vallorbe Tourist Railway at Les Hôpitaux-Neufs

DR 52.8163*	52.5996	-	BMAG	12437	KDL I	1943	2-10-0	-

Great Britain

Bressingham Steam Preservation Co, Norfolk

-	52.5865	NSB 5865	Schichau	3063	KDL I	1944	2-10-0	Now the only Class 52 preserved in GB

DR/DB Nos (pre-computerisation)	DRG No	Other Nos	Builder	Number	ÜK/KDL	Date	Wheels	Comments
Hungary								
Hungarian National Railway Museum at Budapest								
-	52.5156	TE.5156 MAV 520.034	Krenau/Chrzanow	1165	KDL 1	1943	2-10-0	-
GySEV Railway Museum at Fertoboz								
-	52.3535	TE.3535 MAV 520.030	Krauss-Maffei	16661	KDL 1	1943	2-10-0	Sold to GySEV in 1976
Countries of the former Jugoslavia								
Slovenian Railway Museum at Ljubljana								
-	-	HDZ 30.022/ JDZ 33.037	Henschel	27943	KDL 1	1944	2-10-0	Built new for HDZ
-	52.2377	SZD TE.2377 JDZ 33.339	Henschel	27545	KDL 1	1943	2-10-0	-
DB 52.3417	52.3417	JDZ 33.253	Krauss-Maffei	16543	KDL 1	1943	2-10-0	-
Croatian Railway Museum at Zagreb								
DB 52.748	52.748	JDZ 33.237	Henschel	28014	KDL 1	1944	2-10-0	-
-	52.3436	JDZ 33.098	Krauss-Maffei	16562	KDL 1	1943	2-10-0	-
-	52.7684	JDZ 33.161	WLF	17032	KDL 1	1944	2-10-0	-

DR/DB Nos (pre-computerisation)	DRG No	Other Nos	Builder	Number	ÜK/KDL	Date	Wheels	Comments
Serbian Railways Belgrade								
-	52.2802	JDZ 33.087	Henschel	28366	KDL 1	1944	2-10-0	-
Luxembourg								
Luxembourg Railways (CFL)								
-	(42.2718)	CFL 5519	WLF	17615	KDL 3	1948	2-10-0	Never carried its DRG No; initially preserved as CFL No 5513
The Netherlands								
The Veluwse Stream Train at Apeldoorn								
DR 44.1593	44.1593	-	SACM	7849	ÜK	1943	2-10-0	Oil-burning from 1966 to 1982
DR 50.3666*	50.2145	-	Franco-Belge	2567	ÜK	1943	2-10-0	Oil-burning from 1971 as 50.0073
DR 52.8053*	52.2551	-	Henschel	27719	KDL 1	1943	2-10-0	-
-	52.3879	ÖBB 52.3879	WLF	17322	KDL 1	1944	2-10-0	-
South Limbourg Steam Train Co								
DR 52.8160*	52.532	-	BMAG	13099	KDL 1	1943	-	-
Netherlands Steam Foundation at Rotterdam								
-	50.1255	-	Jung	9283	-	1941	2-10-0	-

DR/DB Nos (pre-computerisation)	DRG No	Other Nos	Builder	Number	ÜK/KDL	Date	Wheels	Comments

Norway

Museum at Hamar

-	52.2770	NSB 2770	Henschel	28322	KDL 1	1944	2-10-0	-

Poland

National Railway Museum at Warsaw

-	-	PKP Ty42.120	Chrzanow	1604	KDL 1	1946	2-10-0	Built after war
-	-	PKP Ty43.17	Cegielski	1045	KDL 1	1947	2-10-0	Built after war

The Museum Depot at Chabowka near Krakow

-	52.053	PKP Ty2.29	Henschel	26971	KDL 1	1942	2-10-0	-
-	52.200	PKP Ty2.50	BMAG	12205	KDL 1	1943	2-10-0	-
-	52.1346	PKP Ty2.911	DWM/Cegielski	812	KDL 1	1944	2-10-0	-
-	52.2733	PKP TY2.1184	Henschel	27965	KDL 1	1944	2-10-0	-
-	52.2817	PKP Ty2.953	Henschel	28163	KDL 1	1944	2-10-0	-
-	52.4392	PKP Ty42.19	Chrzanow	1524	KDL 1	1945	2-10-0	Built after war
-	-	PKP Ty43.9	Ceglielski	994	KDL 3	1946	2-10-0	Built after war

The Museum Depot at Jaworzyna

-	86.240	PKP TKt3.16	Schichau	3286	-	1935	2-8-2T	Built before war
-	52.2112	PKP Ty2.81	Henschel	26868	KDL 1	1943	2-10-0	-
-	52.2127	Ty2.223	Henschel	26883	KDL 1	1943	2-10-0	Converted to oil-burning

DR/DB Nos (pre-computerisation)	DRG No	Other Nos	Builder	Number	ÜK/KDL	Date	Wheels	Comments
-	-	Ty42.1	Chrzanow	1506	KDL 1	1945	2-10-0	Built after war

The Museum Depot at Wolsztyn

DR/DB Nos	DRG No	Other Nos	Builder	Number	ÜK/KDL	Date	Wheels	Comments
-	50.451	Ty5.10	Schichau	3413	-	1940	2-10-0	-
-	52.4770	PKP Ty2.406	MBA	13821	KDL 1	1943	2-10-0	-
-	42.1427	Ty43.126/ Ty3.2	Schichau	4448	KDL 3	1944	2-10-0	-
-	-	Ty43.123	Ceglielski	1354	KDL 3	1949	2-10-0	Built after war

'Train for Heaven' Art-Work at Wrocław

DR/DB Nos	DRG No	Other Nos	Builder	Number	ÜK/KDL	Date	Wheels	Comments
-	52.3914	PKP Ty2.1035	MBA	14168	KDL 1	1944	2-10-0	Loco positioned on rails at angle of some 80 degrees from horizontal

Romania

Bucharest

DR/DB Nos	DRG No	Other Nos	Builder	Number	ÜK/KDL	Date	Wheels	Comments
-	-	CFR 150.025	Resita	710	-	1950	2-10-0	Built after war, formerly active at Sibiu depot
-	-	CFR 150.105	Resita	2166	-	1955	2-10-0	1,000th engine constructed at Resita works after war

Sibiu Museum Depot

DR/DB Nos	DRG No	Other Nos	Builder	Number	ÜK/KDL	Date	Wheels	Comments
-	50.3240 52.196	CFR 150.1105	BMAG	12201	ÜK/KDL 1	1943	2-10-0	Built as a ÜK loco but reclassified KDL 1. Formerly active at Sibiu Depot

DR/DB Nos (pre-computerisation)	DRG No	Other Nos	Builder	Number	ÜK/KDL	Date	Wheels	Comments
Open Air Railway Museum at Resita								
-	-	CFR 150.038	Resita	723		1955	2-10-0	Built after war
Switzerland								
Steam Locomotive & Machine Works (DLM) at Schaffhausen								
DR 52.8055*	52.1649	-	SACM	7916	KDL 1	1943	2-10-0	Rebuilt as hi-tech experimental loco
Turkey								
Turkish National Railway Museum at Ankara								
-	-	TCDD 56.504	Henschel	27738	-	1943	2-10-0	Built new for TCDD
Railway Museum at Camlik								
-	52.4862	TCDD 56523	MBA	13926	KDL 1	1943	2-10-0	-
-	(44.1832)	SNCF 150X	Bat	747	ÜK	1945	2-10-0	Built after war; never carried its DRG No; initially SNCF Class 150X. Then TCDD in 1955.
Denzli Station								
-	52.7434	TCDD 56553	WLF	16887	KDL 1	1944	2-10-0	Last of TCDD class numerically

DR/DB Nos (pre-computerisation)	DRG No	Other Nos	Builder	Number	ÜK/KDL	Date	Wheels	Comments
Sultanhisar Station								
-	52.365	TCDD 56512	BLW	15462	KDL I	1943	2-10-0	-
Usak Station								
-	52.7428	TCDD 56547	WLF	16881	KDL I	1944	2-10-0	-
Countries of the former USSR								
Minsk Railway Museum, Belarus								
-	52.5248	TE.5248	Krenau/Chrzanow	1257	KDL I	1943	2-10-0	-
-	52.5653	TE.5653	Schichau	3931	KDL I	1943	2-10-0	-
Brest Railway Museum, Belarus								
-	52.2596	TE.2596	Henschel	27764	KDL I	1943	2-10-0	-
	52.4751	TE.8026	MBA	13802	KDL I	1943	2-10-0	Rebuilt and renumbered in USSR, 1952
Railway Museum at Haapsalu, Estonia								
-	52.3368	TE.3368	Krauss-Maffei	16494	KDL I	1943	2-10-0	-
Railway Museum at Riga, Latvia								
-	52.036	TE.036	Henschel	26954	KDL I	1942	2-10-0	-

DR/DB Nos (pre-computerisation)	DRG No	Other Nos	Builder	Number	ÜK/KDL	Date	Wheels	Comments
Railway museum at Vilnius, Lithuania								
-	52.737	TE.0737	Henschel	27966	KDL I	1944	2-10-0	-
-	52.313	TE.0313	WLF	9685	KDL I	1943	2-10-0	-
-	52.7517	TE.7517	Skoda	1615	KDL I	1944	2-10-0	-
Railway Station at Radviliskis, Lithuania								
-	52.4567	TE.4567	Cegielski	884	KDL I	1945	2-10-0	-
Railway Museum at Kaliningrad, Russia								
-	52.4564	TE.4564	Cegielski	880	KDL I	1945	2-10-0	-
Rizhsky Technical Museum in Moscow, Russia								
-	52.5415	TE.5415	Schichau	3693	KDL I	1943	2-10-0	-
Museum at St Petersburg, Russia								
-	52.6769	TE.6769	WLF	16222	KDL I	1944	2-10-0	-

Table 3
Other selected KDL and HF engines preserved in Great Britain and Europe

Running No	Other No carried	Builder	Works No	Date	KDL	Other Locos	Location
KDL 2: the Czechoslovakian (CSD) Class 534.0 2-10-0s							
CSD 534.027	-	Skoda	267	1923	-	Other	Built before war, now at National Technical Museum, Chomotov
CSD 534.0432	-	CKD	2335	1946	-	Other	Built after war, now at Klatovy Depot for special workings
CSD 534.0456	-	CKD	Unknown	Unknown	-	Other	Built after war, now at Luzna Railway Museum, Czech Republic
CSD 534.0471	-	Skoda	1778	1947	-	Other	Built after war, now at the Slovak Railway Museum, Bratislava
KDL 4: 0-8-0 industrial tanks							
-	DFS 4	BMAG	9963	1930	-	Other	Built before war, now at Railway Museum at Ebermannstadt, Germany
-	DME 184	Henschel	25657	1946	-	Other	Built after war, now at Railway Museum at Darnstadt-Kranichstein, Germany
KDL 8: 0-4-0 industrial tanks							
-	SBS No 1	WLF	16111	1944	KDL 8	-	OeGEG Railway Museum, Amplwang, Austria
-	SBS No 2	WLF	17324	1944	KDL 8	-	Private Railway Museum, Hermeskeil, Germany
-	SBS No 4	WLF	17323	1944	KDL 8	-	Transport Museum, Prora, Germany
KDL 10: 900mm gauge 0-4-0 construction tanks							
-	Unknown	Krauss-Maffei	17420	1944	KDL 10	-	Kohlebebanen at Haselbach, Germany

Running No	Other No carried	Builder	Works No	Date	KDL	Other Locos	Location
KDL 11: 600/750/760mm industrial tanks, Heeresfeldbahnlokomotiven (HF Class 160D)							
ÖBB 699.101	HF 2817	Franco-Belge	2817	1944	KDL 11	-	Gurktalbahn, Austria
ÖBB 699.01	HF 2818	Franco-Belge	2818	1944	KDL 11	-	Club 760, Taurachbahn, Austria
ÖBB 699.102	HF 2821	Franco-Belge	2821	1944	KDL 11	-	OeGEG Stytalbahn, Austria
ÖBB 699.02	HF 2822	Franco-Belge	2822	1944	KDL 11	-	Technical Museum, Berlin
CFCD 10	HF 2836	Franco-Belge	2836	1945	KDL 11	-	Built after war, now at Le P'tit train de la Haure Somme, Dompiere, France
AMTP 4-12	HF 2843	Franco-Belge	2843	1945	KDL 11	-	Built after war, now at Transport Museum, Pithivier, France
CFBB 4-13	HF 2844	Franco-Belge	2844	1945	KDL 11	-	Built after war, now owned by APEMVE, Saint Germain d'Arce, France
-	HF 2845	Franco-Belge	2845	1945	KDL 11	-	Built after war, now at Tourist Railway of the Doller Valley, Burhaupt-le-Haut, France; converted to standard gauge
W&LR 10	HF 2855	Franco-Belge	2855	1945	KDL 11	-	Built after war, now running as *Sir Drefaldwyn*, Welshpool & Llanfair Railway, Wales
KDL 13: 600/750mm 'Riesa'-type construction tanks (HF Class 70PS)							
PKP T3-1043	-	Budich, Wroclau	931	1943	KDL 13	-	Narrow Gauge Museum, Sochaczew, Poland
-	DDM 3	Budich, Wroclau	1028	1944	KDL 13	-	Railway Museum, Neuenmarkt-Wirsberg, Germany
-	DDM 2	Budich, Wroclau	1029	1944	KDL 13	-	Field Army Railway Museum, Fortuna, Germany
-	HF 5026	Henschel	24011	1939	-	-	Army Railway Museum, Frankfurt-Main, Germany
-	DKBM 2	Henschel	28470	1948	-	-	Built after war, now at Narrow Gauge Steam Museum, Mühlenstroth, Germany
-	-	Henschel	28514	1949	-	-	Built after war, now at Technical Museum, Berlin

Running No	Other No carried	Builder	Works No	Date	KDL	Other Locos	Location
Heeresfeldbahnlokomotiven 600/750mm gauge 0-6-0 tanks (HF Class 110C)							
-	HF 25982	Henschel	25982	1941	-	-	Narrow Gauge Museum, Gutersloh Germany
DR 99.4652	HF 25983	Henschel	25983	1941	-	-	RUKB, Putbus, Germany
-	HF 11017	Krenau/Chrzanow	945	1941	-	-	Railway Museum, Lavassaare, Estonia
HF 750/760mm gauge 0-10-0 tanks (HF Class 210 E)							
-	HF 191	BLW/Borsig	14806	1939	-	-	Club 760, Taurachbahn, Austria
-	HF 24755	Henschel	26466	1944	-	-	Thought to be at Bruchhausen-Vilsen Museum, Germany

Table 4
Selected MoS engines preserved in Great Britain and elsewhere

Engines ordered by MoS for WD and REC	WD Nos/names	Subsequent owners	Other pre-preservation Nos/names	Builder	Works No	Date	Location and name(s) in preservation
Stanier 8F 2-8-0							
WD	340	TCDD	45168	NB	24640	1940	Izmit, Turkey
WD	341	TCDD	45166	NB	24641	1941	Be'er Shiva, Israel
WD	348	TCDD	45160	NB	24648	1940	GCR Ruddington, Nottinghamshire
WD	353	TCDD	45165	NB	24653	1940	Alaseher Station, Turkey
WD	522	TCDD	45161	NB	24670	1941	Camlik, Turkey
WD	554	TCDD	45170	NB	24755	1942	Bo'ness & Kinneil Rly, West Lothian
REC	-	LMS/BR	8431/48431	Swindon	-	1944	K&WVR, West Yorkshire

Engines ordered by MoS for WD and REC	WD Nos/names	Subsequent owners	Other pre-preservation Nos/names	Builder	Works No	Date	Location and name(s) in preservation
REC	-	LMS/BR	8518/48518	Doncaster	-	1944	GWS, Didcot, Oxfordshire; boiler used in new 'County Project'
REC	-	LMS/BR	8624/48624	Ashford	-	1943	GCR Ruddington, Nottinghamshire
WD	(70)397/500	LMS/BR	8233/48773	NB	24607	1940	SVR, Highley, Shropshire

Hunslet 'Austerity' saddle tanks (ex-WD engines only)

Engines ordered by MoS for WD and REC	WD Nos/names	Subsequent owners	Other pre-preservation Nos/names	Builder	Works No	Date	Location and name(s) in preservation
WD	(7)1463	LNER/BR/NCB	8078/68078/L2	AB	2212	1946	Hope Farm, Sellindge, Kent
WD	(7)1466	LNER/BR/NCB	8077/68077/14	AB	2215	1946	Spa Valley Railway, Tunbridge Wells, Kent
WD	(7)1480	NCB		RSH	7289	1945	Tyseley Locomotive Works, Birmingham *Fred*
WD	(7)1499	MoF&P/NCB		HC	1776	1944	Bryn Engineering Services, Bolton, Lancashire *Harry*
WD	(7)1505/118 *Brussels*	NCB	-	HC	1782	1945	K&WVR, West Yorkshire *Brussels*
WD	(7)1515	NCB	-	RSH	7169	1944	Pontypool & Blaenavon Railway, Gwent *MECH. NAVVIES LTD*
WD	(7)1516	NCB	-	RSH	7170	1944	Gwili Railway, Carmarthenshire *Welsh Guardsman*
WD	(7)1529/165	Wemyss Private Railway	15	AB	2183	1943	Sharpness Docks, Gloucestershire
WD	(7)5008	MoF&P/NCB	-	HE	2857	1943	Nene Valley Railway, Peterborough, Cambridgeshire *Swiftsure*
WD	(7)5015	MoF&P/NCB	-	HE	2864	1943	Aln Valley Railway, Northumberland
WD	(7)5019/118	NCB	-	HE	2868, r/b 3883	1944/1963	Midland Railway Centre, Butterley *Lord Phil*
WD	(7)5024	NS	8811/LV 12	HE	2873	1943	Stoom Stichting, Netherlands
WD	(7)5031/101	NCB	17	HE	2880	1943	Bo'ness & Kinneil Railway, West Lothian

Engines ordered by MoS for WD and REC	WD Nos/ names	Subsequent owners	Other pre-preservation Nos/names	Builder	Works No	Date	Location and name(s) in preservation
WD	(7)5041/107 *Foggia*	NCB	*Maureen*	HE	2890, r/b 3882	1943/1960	East Lancashire Railway, Bury, Lancashire Converted to 0-6-0 tender engine *Douglas*
WD	(7)5050	NCB	-	RSH	7086	1943	Bryn Engineering Services, Bolton, Lancashire *Rolvenden/Norman*
WD	(7)5061	MoF&P/NCB	-	RSH	7097	1943	Strathspey Railway, Aviemore, Highland *Cairngorm*
WD	(7)5062	MoF&P/NCB+C31		RSH	7098	1943	Tanfield Railway, Co Durham
WD	(7)5091	NCB	7 *Robert*	HC	1752	1943	GCR Ruddington, Nottinghamshire (as 68067)
WD	(7)5105	NS/Laura Mine	8815/LV 15	HE	3155	1944	Ribble Railway, Preston, Lancashire *Walkden*
WD	(7)5113/132 *Sapper* (1)	NCB	*Alison*	HE	3163	1944	Avon Valley Railway, Bitton, Gloucestershire *Sapper*
WD	(7)5115	NS/Laura Mine	8826/LV 16	HE	3165	1944	Zuid-Limburgse Stoomtrein, Limburg, Netherlands
WD	(7)5118/134 *King Feisal*	NCB	S134	HE	3168	1944	Embsay & Bolton Abbey Steam Railway, North Yorkshire *Wheldale*
WD	(7)5142/140	NCB	-	HE	3193 r/b 3887	1944 r/b 1964	Bressingham Live Steam Museum, Norfolk *Norfolk Regiment*
WD	(7)5170	MoF&P/NCB	-	WGB	2758	1944	Cefn Coed Colliery Mining Museum, Port Talbot *Cefn Coed Colliery*
WD	(7)5186/150 *Royal Pioneer*	NCB	-	RSH	7136	1944	Peak Rail, Rowsley, Derbyshire *Royal Pioneer*
WD	(7)5189/15 *Rennes*	NCB	8	RSH	7139	1944	Dean Forest Railway, Gloucestershire *Rennes*
WD	(7)5256	MoF&P/NCB	WD 1/*Gamma*	WGB	2777	1945	Bo'ness & Kinneil Railway, West Lothian
WD	(7)5282/81 *Insein*	NCB	*Haulwen*	VF	5272	1945	Gwili Railway, Bronwydd Arms, Carmarthenshire
WD	(7)5300	CFT	3.55	VF	5290	1945	Railway Museum, Farhat Hached, Tunisia

Engines ordered by MoS for WD and REC	WD Nos/ names	Subsequent owners	Other pre-preservation Nos/names	Builder	Works No	Date	Location and name(s) in preservation
WD	(7)5319	NCB	72	VF	5309	1945	Peak Rail, Rowsley, Derbyshire
WD	190/90	-	-	HE	3790	1953	Colne Valley Railway, Castle Hedingham, Essex *Castle Hedingham*
WD	191/91 *Black Knight*	-	-	HE	3791	1953	K&ESR, Rolvenden, Kent *Holman F. Stephens*
WD	192/92 *Waggoner*	-	-	HE	3792	1953	Isle of Wight Steam Railway, Havenstreet *Waggoner*
WD	193/93	-	-	HE	3793	1953	Ribble Railway, Preston, Lancashire *Shropshire*
WD	194/94	-	-	HE	3794	1953	Embsay & Bolton Abbey Steam Railway, North Yorkshire *Cumbria*
WD	196 *Errol Lonsdale*	-	-	HE	3796	1953	Stoomcentrum, Maldegem, Belgium *Errol Lonsdale*
WD	197 *Sapper* (2)	-	-	HE	3797	1953	K&ESR, Rolvenden, Kent *Northiam*
WD	198/98 *Royal Engineer*	-	-	HE	3798	1953	Isle of Wight Steam Railway, Havenstreet *Royal Engineer*
WD	200/95	-	-	HE	3800	1953	Colne Valley Railway, Castle Hedingham, Essex *William H. Austen*

WD 'Austerity' 2-8-0

WD	(7)9257	NS/SJ	4464/1931	VF	5200	1945	K&WVR (as 90733)

WD 'Austerity' 2-10-0s

WD	(7)3651/ 600 *Gordon*	-	-	NB	25437	1943	SVR, Highley, Shropshire *Gordon*
WD	(7)3652	HSR	Lb951	NB	25438	1943	North Norfolk Railway, Sheringham (as 90775) *Norfolk Regiment*

Engines ordered by MoS for WD and REC	WD Nos/ names	Subsequent owners	Other pre-preservation Nos/names	Builder	Works No	Date	Location and name(s) in preservation
WD	(7)3656	HSR	Lb955	NB	25442	1943	Thessalonica OSE Depot, Greece (dumped but intact in 2019)
WD	(7)3672	HSR	Lb960	NB	25458	1943	NYMR, Grosmont *Dame Vera Lynn*
WD	(7)3677	HSR	Lb962	NB	25463	1943	Drama OSE Depot, Greece (intact in 2019)
WD	(7)3682	HSR	Lb964	NB	25468	1944	Thessalonica OSE Depot, Greece
WD	(7)3699	HSR	Lb958	NB	25445	1943	Tithorea OSE Depot, Greece
WD	(7)3755 *Longmoor*	NS	5085	NB	25601	1944	Dutch National Railway Museum, Utrecht *Longmoor*

Southern Railway Bulleid Class Q1 0-6-0

SR	-	-	C1/33001	SR Brighton	-	1942	National Railway Museum, York

Table 4a
Other Stanier 8Fs and 'Austerity' 0-6-0 saddle tanks preserved in Great Britain

Engines ordered by	Subsequent owners	Nos carried	NCB name	Builder	Works No	Date	Location and other name(s) in preservation

Stanier 8F 2-8-0s

LMS	BR	8151/48151	-	Crewe	-	1942	West Coast Railways, Carnforth, Lancashire
LMS	BR	8173/48173	-	Crewe	-	1942	Churnet Valley Railway, Staffordshire
LMS	BR	8305/48305	-	Crewe	-	1943	GCR, Loughborough, Leicestershire

Engines ordered by	Subsequent owners	Nos carried	NCB name	Builder	Works No	Date	Location and other name(s) in preservation
Hunslet 'Austerity' saddle tanks							
NCB	-	60	-	HE	3686	1948	Aln Valley Railway, Northumberland
NCB	-	-	Whiston	HE	3694	1950	Foxfield Railway, Staffordshire
NCB	-	-	-	HE	3698	1950	Lakeside & Haverthwaite Railway, Cumbria *Repulse*
NCB	-	8	Bickershaw	HE	3776	1952	Bryn Engineering Services, Bolton, Lancashire *Sir Robert Peel/Warspite*
NCB	-	9	-	HE	3777	1952	Llangollen Railway, Denbighshire *Josiah Wedgwood*
NCB	-	-	-	HE	3781	1952	Mid-Hants Railway (*Linda*, now as blue side-tank *Thomas*)
NCB	-	69	-	HE	3785	1953	Embsay & Bolton Abbey Steam Railway, North Yorkshire
NCB	-	1	Monkton	HE	3788	1953	Embsay & Bolton Abbey Steam Railway, North Yorkshire
NCB	-	4	-	HE	3806	1953	Dean Forest Railway, Gloucestershire *G. B. Keeling/ Wilbert, Rev W. Awdry*
NCB	-	18	-	HE	3809	1953	GCR, Loughborough
NCB	-	-	Glendower	HE	3810	1954	South Devon Railway, Buckfastleigh
NCB	-	19	-	HE	3818	1954	Bo'ness & Kinneil Railway, West Lothian
NCB	-	-	Warrior	HE	3823	1954	Dean Forest Railway, Gloucestershire
NCB	-	9	-	HE	3825	1954	Stainmore Railway, Kirkby Stephen, Cumbria
NCB	-	-	-	HE	3829	1955	Gwili Railway, Bronwydd Arms, Carmarthenshire
NCB	-	5/21	-	HE	3837	1955	Bo'ness & Kinneil Railway, West Lothian
NCB	-	-	-	HE	3839	1956	Foxfield Railway, Blythe Bridge, Staffordshire *Wimblebury*
NCB	-	-	Pamela	HE	3840	1956	Gwili Railway, Bronwydd Arms, Carmarthenshire
NCB	-	22	-	HE	3846	1956	Appleby Frodingham Railway, Scunthorpe, North Lincolnshire

Engines ordered by	Subsequent owners	Nos carried	NCB name	Builder	Works No	Date	Location and other name(s) in preservation
Stewart & Lloyds Minerals	-	-	*Juno*	HE	3850	1958	NRM, Shildon, Co Durham
NCB	-	65	-	HE	3889	1964	Dean Forest Railway, Lydney, Gloucestershire
NCB	-	66	-	HE	3890	1964	Buckingham Railway Centre, Quainton Road, Buckinghamshire

Table 5
Selected USATC engines preserved in Great Britain and Europe

USATC No	Other Nos carried	Builder	Works No	Date	Location
USATC Class S100 'USA' 0-6-0 tanks (ex-USATC engines only)					
1310	HSR Da 61	Dav	2481	1943	Thessalonica Depot OSE, Greece (dumped but intact in 2017)
1396	JZ 62.084	Porter	7510	1942	Gracac Station, Croatia
1400	JZ 62.086	Porter	7514	1942	Restaurant at Pancevo, Serbia
1959	BR 30064	VIW	4432	1943	Bluebell Railway, Sheffield Park, West Sussex
1968	BR 30065, DS237 Maunsell	VIW	4441	1943	Kent & East Sussex Railway, Rolvenden, Kent
1960	BR 30070, DS 238 Wainwright	VIW	4433	1943	K&ESR, Rolvenden, Kent
1973	BR 30072	VIW	4446	1943	K&WVR, Haworth, West Yorkshire
1987	HSR Da 65	VIW	4460	1943	Tithorea Depot OSE, Greece (dumped but intact in 2011)
1999	HSR Da 55	VIW	4472	1943	Thessalonica Railway Museum, Greece
4383	SNCF 030TU22	Dav	2532	1943	Longueville Railway Museum, France
4389	Orange Nassau Mine, Netherlands No ON-27	Dav	2538	1943	Stoomtrein Goes Borsele, Zeeland, Netherlands

USATC No	Other Nos carried	Builder	Works No	Date	Location
5042	JZ 62.046	Porter	7584	1943	Haut Quercy Tourist Railway, Martel, France (as 030TU46)
6008	HSR Da 63	Dav	2597	1943	Thessalonica Railway Museum, Greece
6022	JZ 62.054	Dav	2611	1944	Croatian National Railway Museum, Zagreb
6102	SNCF 030TU13	Porter	7682	1944	CF de la Suisse Normande, St Pierre du Regard
6176	JZ 62.074	VIW	4546	1944	Ruma Station, Serbia

USATC Class S160 2-8-0s

USATC No	Other Nos carried	Builder	Works No	Date	Location
1631	MAV 411.388	ALCO Montreal, Canada	70284	1942	GCR Ruddington, Nottinghamshire
2206	HSR THg 525	LIMA	8212	1943	Thessalonica OSE Depot, Greece (dumped but intact in 2019)
2226	HSR THg 535	LIMA	8232	1943	Thessalonica OSE Depot, Greece (dumped but intact in 2019)
2253	PKP TR 203-288	BLW	69496	1943	Dartmouth Steam Railway. Devon *Omaha*
2364	MAV 411.337	BLW	69621	1943	GCR Ruddington, Nottinghamshire
2438	PKP Tr203-296	ALCO	70787	1943	Jaworzyna Railway Museum, Poland
2524	TCDD 45172	LIMA	8341	1943	Camil Railway Museum, Turkey
2781	MAV 411.264	LIMA	8434	1943	Hatvan Station, Budapest
2879	TCDD 45174	ALCO	71076	1943	Ankara Railway Museum, Turkey
3278	FS 736.073, HSR THg 575	ALCO Montreal, Canada	71533	1944	Tyseley Locomotive Works, Birmingham *Franklin D. Roosevelt*
3292	FS 736.083	ALCO	71555	1944	Piedmont Railway Museum, Savigliano, Italy
3299	FS 736.090, HSR THg 576	ALCO	71554	1944	Tithorea Depot OSE, Greece (dumped but intact in 2011)
3324	FS 736.114	ALCO	71579	1944	Pietrarsa San Giorgio Railway Museum, Naples, Italy
3450	MAV 411.118	BLW	70407	1944	Hungarian Heritage Railway Park, Budapest
3524	HSR THg 532	BLW	70481	1944	Thessalonica Depot OSE, Greece (dumped but intact in 2019)
3699	FS 736.208	LIMA	8612	1944	Milan Smistamento Depot, Italy (dumped but intact in 2019)

USATC No	Other Nos carried	Builder	Works No	Date	Location
5801	PKP Tr203-451	LIMA	8739	1945	Polish National Railway Museum, Warsaw
5820	PKP Tr203-474	LIMA	8758	1945	K&WVR, West Yorkshire
5164	PKP Tr201-54	LIMA	8823	1945	Jaworzyna Railway Museum, Poland
6046	MAV 411.144	BLW	72080	1945	Churnet Valley Railway, Cheddleton, Staffordshire
6056	MAV 411.358	BLW	72090	1945	Hegyeshalom Station, Budapest

Table 5a
USATC designed engines, now preserved in Great Britain

USATC Class S100 engines built in Jugoslavia					
-	JZ 62.521 (Industrial)	DD	521	1955	Shillingstone Station Project, Dorset (to be 30076)
-	JZ 62.669 (Industrial)	DD	669	1960	Shillingstone Station Project, Dorset (as 30075)
USATC Class S160 2-8-0 that worked in China					
5197	KD 6.463	LIMA	8856	1945	Churnet Valley Railway, Cheddleton, Staffordshire

Silver Link Silk Editions

In March 2014 we introduced the first of our Silver Link Silk Editions, which will feature a silver, gold or green silk style bookmark (the use of such silks dates back to the reign of Elizabeth I). Printed on high quality gloss art paper, these sewn hardcover volumes also feature head and tail bands. Such quality and tradition will be much welcomed by today's discerning print book readers.

Further Silk Edition volumes will be made available from time to time and details will be shown on our website:
www.nostalgiacollection.com

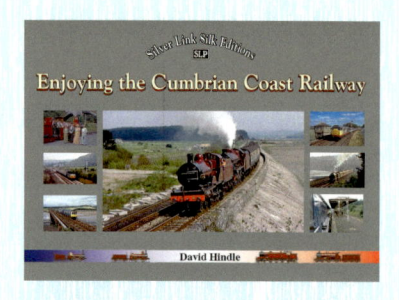

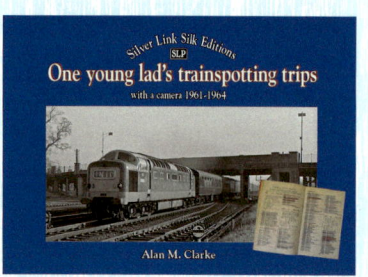

Further information

Silver Link and Past & Present titles are available while stocks last through bookshops, preserved railways and many heritage sites throughout the UK.

Further details can be found on our website:
www.nostalgiacollection.com

Our latest catalogue is also available on request by writing to us at the address shown on the title page of this volume or by emailing your request to:

sohara@mortons.co.uk

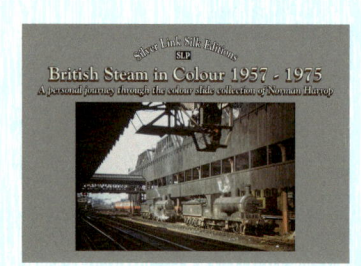

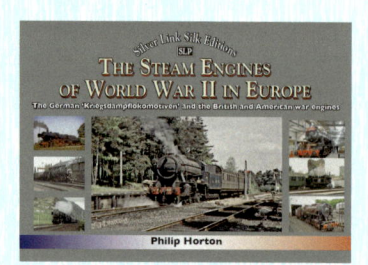

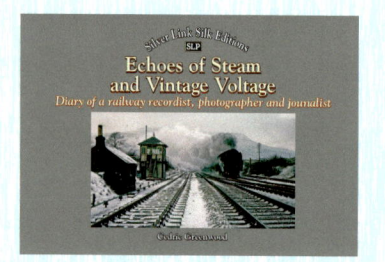